Advance Praise for S

The modern world depends on a fe
GPS, and, of course, the World Wide Web. And all of those networks rely on the mother network: the electric grid. In *Shorting the Grid*, Meredith Angwin provides an enormously valuable, clear, and succinct explanation of our most important network. She shows how it works, why it's vulnerable, and why we should be concerned about what she lyrically calls the "angelic miracle of the power grid." If you care about the future of our increasingly electrified world, buy this book and read it.

— Robert Bryce, author of *A Question of Power: Electricity and the Wealth of Nations*

An eye-opening exposé of our grid's vulnerabilities. The "deregulated" grid is highly political, secretive, overly complex, and unable to meet public needs like reliability, affordability, and low pollution. If you take for granted that the lights go on when you flip a switch, this book may blow your mind.

— Joshua S. Goldstein, author of *A Bright Future: How Some Countries Have Solved Climate Change and the Rest Can Follow*

Shorting the Grid is full of sharp writing and engaging stories about the most hidden part of our grid—how grid-level decisions are made. Readers will come away with an understanding of the Grid not just as a machine, but as a web of people and policies that shape the rules on how energy is made, paid for, moved, and used.

— Dan Nott

Dan Nott is the artist and author of *Hidden Systems*, a forthcoming graphic non-fiction book about major types of infrastructure.

The North American electric grid is the largest machine in the world. Electricity is key to modern civilization, providing energy in its most useful form, powering industry, commerce, and living. Future blackouts will cost the U.S. economy $2 billion an hour. However, power-generating utilities don't run the grid anymore. Other empowered stakeholders pursue diverging objectives, which means that prices rise and grid reliability drops.

Why? It's complicated. Reading Angwin's book is like chatting with an expert who helps you understand the underlying engineering, finances, and policies creating the risks. Her narrative moves back and forth between insightful overviews and specific examples. The book covers many grid attributes, suggesting realistic conclusions without ideological advocacy.

This book is a must-read for anyone suggesting any improvement in our electricity supply, for academics trying to conduct analyses, for businesses needing continuous economic energy, for legislative and regulatory staffers, and energy-concerned citizens.

— Dr. Robert Hargraves, Author of *Thorium, Energy Cheaper than Coal* and co-founder of ThorCon Power

The electric power grid is the most important machine in the modern world. It is poised to become even more important as electricity is tapped to take over from fossil fuels in the fight against climate change. The grid will soon be all that it is today plus our gas station and home heating, not to mention a huge source of revenue for city governments. So it is remarkable that, at exactly this critical moment, grid management and planning have descended into a morass of secret cross purposes with nobody in charge and profiteers running rampant. How can we fix it? To fix the grid we have to know it, and the very best way to know it is to read Meredith Angwin's timely and alarming new book *Shorting the Grid*.

— Steve Aplin, blogger at *Canadian Energy Issues*

The National Academy of Engineering describes the U.S. power grid as the "supreme engineering achievement of the 20th century." Its user interface has customer-facing simplicity that hides a system of massive complexity. The grid's engineering excellence has allowed almost all Americans to simply assume its future existence.

But Meredith Angwin is an unusually perceptive maven who digs deeply when initial answers don't make sense. She's learned that the marvelous machine that is the foundation of modern living is being threatened by political and financially motivated maneuvers.

Shorting the Grid reveals reasons why we must pay more attention to grid governance and the potential of poor decisions to override technical successes.

— Rod Adams: Blogger at *Atomic Insights,* Managing Member at Nucleation Capital LP

Shorting the Grid reveals what may be the ultimate in Regulatory Capture: so-called "deregulated" electrical markets. In this book, author Meredith Angwin exposes the "deregulated" energy markets as a frightening example of the "Golden Rule" at play. Utilities, generators, and other moneyed interests have the Gold, so they get to make the Rules for the electrical grid that we all rely upon. One troubling consequence: nobody is responsible for ensuring there's enough electricity on the grid to meet demand.

With this eye-popping tour de force, author Meredith Angwin paints an infuriating portrait of fat-cat insiders making backroom deals to sweeten their profits even at the expense of the reliability of the grid they are supposedly entrusted with operating. Meanwhile, the public is left out of the loop and out in the cold.

Essential reading for anyone interested in clean energy policy and decarbonization.

— David Schumacher, Director of *The New Fire*

SHORTING
THE GRID

SHORTING THE GRID

THE HIDDEN FRAGILITY OF OUR ELECTRIC GRID

MEREDITH ANGWIN

CARNOT COMMUNICATIONS

Shorting the Grid: The Hidden Fragility of Our Electric Grid

For information about this title or to order other books and/or electronic media, contact the publisher:
Carnot Communications
PO Box 741
Wilder, VT 05088
http://www.meredithangwin.com
mjangwin@gmail.com

Carnot Communications publications are available at special discounts for bulk purchases in the U.S. by corporations, institutions, and other organizations. For more information, please contact the author at mjangwin@gmail.com.

All pictures not otherwise credited are credited to Meredith Angwin.

Hardcover ISBN: 978-0-9891190-8-5
Paperback ISBN: 978-1-7353580-0-0
Ebook ISBN: 978-0-9891190-9-2

Cover and interior design: 1106 Design

Published in the United States of America. Application submitted for Library of Congress registration.

Dedicated to George, my partner in life, and to our children and their families

TABLE OF CONTENTS

TABLE OF FIGURES

ANGELIC MIRACLES AND EASY PROBLEMS

CHAPTER 1

THE BIG SHORT

The Grid and I

I JUST FINISHED REREADING a frightening book, *The Big Short,*[1] which describes the financial meltdown of 2008 and the few people who saw it coming. The book followed how these insightful people placed bets ("shorts") on the fall of complex credit instruments, including credit-default swaps. Some of the people who bet against the credit instruments made hundreds of millions of dollars.

The credit system had little oversight, and it propped up an overheated housing market. "Liar loans" were common: in these loans, the borrower did not have to include any documentation. The overheated housing market, in turn, propped up the wider economy. Shortly after the credit system crashed, everything crashed.

In today's grid governance, I see more parallels with the 2007 financial system than I would like to see.

In 2007, people bought obscure credit instruments, which themselves were based on credit "tranches," and these tranches were backed by "liars mortgages," with no down payment and no credit check. And yet, it was assumed that all these tranches were basically secure. The best way to get rich was to buy the bad tranches because there was no danger of massive default on any tranche. The bad tranches had higher interest rates and were more highly profitable.

In the old days, you could loan mortgage money to cardiac specialist Dr. Jim and his wife, Diane, a schoolteacher. This couple had a sizable down payment, high income, and high credit ratings. In 2007, loaning money to a strawberry picker who had no down payment, little income, and no credit rating was an acceptable way to wealth. The quality of the loan didn't matter anymore. The world of credit was completely upside down, and trouble was sure to come. It did.

The financial activities described in *The Big Short* have many parallels with the current state of much of the power grid in America. In the old days, regulatory bodies wanted to see a grid with reliable power plants and, hopefully, plants that used several different types of fuels. A varied grid meant that, if one fuel had shortages or rose in price, the grid would still be stable, and cost would remain relatively stable.

In current grid governance, none of these things matter. In many areas, power plants that make steady, reliable power can't make a profit. Several large utilities are trying to sell or shut down their nuclear, gas, and coal plants in these areas. These utilities plan to operate plants only in other parts of the country.

Utilities are leaving the Regional Transmission Organization (RTO) areas. These areas have auctions for electricity (and for many other facets of electricity supply). They are also the areas about which you can see occasional news stories about grid-wide problems. A typical news story would be "such and such a grid operator warns of possible electricity shortfalls this summer."

In many areas of the country, but especially in RTO areas, power installations that can operate only intermittently, such as solar and wind installations, are the sure bet for becoming wealthy. In the mortgage situation, the intrinsic value of the mortgage didn't matter. In the RTO area, the value of the power produced doesn't matter. As a matter of fact, less-valuable power is more profitable. Trouble is sure to come, and it is on its way. In these areas, we are on our way to an expensive and fragile grid.

Natural gas is inexpensive. This is a good thing. However, the constant statements that other types of plants "can't compete" with natural gas is not about the plants: it is a consequence of insider decisions on the grid. That is why this book is an exposé.

Nobody has the responsibility

In the RTO areas, no group or agency has the responsibility for grid reliability. This agency can do a little of this, and that agency can do a little of that, but no agency is charged with ensuring reliable power. No agency is in charge of ensuring that there are enough power plants and power lines to keep the grid operating.

In RTO areas, the grid is becoming more fragile and more expensive. Fragility is the most dangerous problem. In the near future, "rolling blackouts" may become common in many RTO areas. This book is about why this will happen and what we can do to prevent those blackouts.

What about the "free market," which could conceivably use its invisible hand to bring reliable electricity to the customers? There is no free market. There are false markets, ruled by insider decisions.

In my opinion, a grid meltdown is coming. Reliable power will become part of the Good Old Days that parents tell their children about.

Unlike the heroes of *The Big Short,* I am not in a position to place some sort of bet that will make me rich. Instead, I will write about the grid and where it is headed. At the end, I will include a few ideas about how grownups (mere "ratepayers") can step up to the plate and take charge.

The ratepayers are the true stakeholders of the grid: we pay for it. It is time that our voices be heard.

The insiders of RTO

THE ELECTRICITY MARKETS in the RTO areas serve nobody well. First of all, they aren't markets. Most types of plants are constantly on the search for their "missing money": the RTO regulations do not allow them to recapture even their costs. Incentives of various types lead to fragile grids, and nobody is the wiser. Huge decisions are made in closed rooms with only insiders (called "stakeholders") present, and the press is often not allowed.

I have come to hate the term "stakeholder." Insiders do the same back-room dealing as the most cynical of big-city bosses in the old days, but this time, it is done in groups called "participant committees." The people in the committees are called "stakeholders" and other such euphemisms for "insiders."

Tacitus described Roman conquest as "They make a desert, and they call it peace." I would describe the RTO areas as "They reward their favorites, and they call it a market."

As a matter of fact, I almost did not write this book. My earlier two books (*Voices for Vermont Yankee*[2] and *Campaigning for Clean Air: Strategies for Pro-Nuclear Advocacy*[3]) were comparatively straightforward: one was people's statements in favor of our local power plant, and the other was a compendium of ideas on how to be an advocate for nuclear power. I can guarantee that, if you follow the advice in the *Campaigning* book, you will make a difference. You may or may not keep your local plant from being shut down by anti-nuclear activists, but you may save it from premature closing, and you will make it impossible for anti-nuclear activists to completely dominate the conversation.

With this book, I felt differently. The grid in RTO areas has a closed-door governance. In many cases, you can't find out what is going on and what decisions are being made. Why should I write a book saying: "This is how you are going to get beaten up. Get used to it"?

Finally, I decided that putting a spotlight on the grid was worthwhile. The first step toward "not being beaten up" is to notice where the attack is coming from or likely to come from. This book shows what happens on the grid and what to watch out for. How to at least find out about important dockets and why they are important. How to write letters about those dockets. And so forth.

Most of the book is factual, but some of it is about how to advocate for change.

In the RTO areas, it's hard to make any difference at all. But if you don't try, nothing will happen. You or your children will wake up to an expensive, unreliable grid. The grid's fragility will have economic and health consequences. And you won't have even tried to stop it.

CHAPTER 2

LOST IN THE THICKET

Once a maven

HOW DID I GET INTERESTED in the grid? As with many stories, it started with arrogance. In particular, it started with me thinking I understood the basics of the electric industry. And then .... I read a headline about our local grid and a local plant. I had no idea what the article was talking about. I realized that I understood nothing.

At that point, for many years, I had thought I was a utility maven. (In Yiddish, *maven* is an ironic term for "a person with deep knowledge.") I had worked in various facets of the utility industry for years. I had even worked on improving electric transmission and distribution (improving the grid). However, most of my work was research about *generating* electricity, not *transmitting* it. I did research on corrosion control for hydro plant penstocks, natural gas pipelines, nuclear plants, and geothermal plants. Besides corrosion control, I researched pollution control (NOx and SOx) for coal and gas plants, and hydrogen sulfide pollution control for geothermal plants.

I'm a chemist, which means a materials person, so all my research was on materials problems, including my work on transmission issues. In that case, I worked on protecting underground electric equipment from thermal runaway. I am the co-inventor on a patent on this subject, for what that is worth. It seems to have gathered some citations, so maybe it was worth something.[4] I have other patents on pollution control for fossil plants.

To do research in so many fields, I had to learn the basics of many different technologies. As a materials person, I worked in almost every phase of the utility industry. After some years, I no longer worked in a lab, but in an office, planning projects and writing papers. For this work, I needed to know about how utilities were regulated (the Clean Air Act, for example) and how they interacted with public utility commissions and other regulators.

So, I was a maven. After many years in the utility industry, I felt I knew the basics. If I read an article about the utility industry, I could generally explain the gist of that article to a non-industry person (for example, my husband). When my husband asked me if we should buy Enron stock (it seemed to be a big deal everywhere at the time), I answered: "No. Enron doesn't have a business model I can understand. I just don't get their 'trading room.'" (The room was a fraud, and I was not surprised when Enron failed.) I was the go-to person for my friends when they wanted to know something about a utility.

I was definitely a maven.

I was knowledgeable about many areas of the utility industry, and I had become (and have remained) very pronuclear. In January 2010, I started a blog, *Yes Vermont Yankee*, about the importance of keeping our local nuclear plant operating.[5]

In August 2010, I saw a newspaper headline about Vermont Yankee. The headline was a shock to me. I did not understand it at all. The headline was: "Grid Operator: Vermont Yankee Not Allowed to De-List from Forward Capacity Auction."[6]

Vermont Yankee was "not allowed to de-list from the forward capacity auction." My reaction was: huh? I understand each word, but what are they saying? What does the entire sentence mean?

I realized that, in terms of how the grid works, I was far from a maven. I couldn't even decipher a headline.

I have been working on my project of understanding the grid for ten years now. It is hard work. How the grid is managed is not transparent, and it is not intuitive. People who work in the utilities do not necessarily understand it. Often, people who work in the regulatory bodies can't answer my questions and refer me to someone else who can answer them. I am not alone in my confusion. Legislators can get quite annoyed at the lack of transparency.

The layers of regulation in the supposedly deregulated RTO areas are mind-boggling. You could hide an elephant in these regulations.

That is a problem. The grid is too important to be left to the control of insiders.

The book and the thicket

I THOUGHT ABOUT ALL the regulations in the RTO areas and how all the RTO areas are different. At that point, I became discouraged about even writing a book about the grid. It could easily turn into such a long, fiercely complicated book that nobody would read it. (I own several grid books of this type. They are mostly meant for lawyers.)

After nine years of study and involvement in industry committees, I have learned the basics. I could name this book: *Grid Governance, Simplified.*

But why write about grid governance, anyway?

I am writing because we need to make major changes in grid governance. We need sunshine laws and new policies. However, changes won't happen unless we know about the current grid. That is why I have named this book *Shorting the Grid.* Our current policies are short-circuiting what should be a reliable grid, which is steadily becoming more dependent on a single fuel, which is delivered "just in time" to be used. As in my other books, this book is mainly a "call to action," not an expository text.

Unfortunately, the layers of regulation on the grid make it very hard to write a "call to action" without a fair amount of exposition. The rules on the grid are a complex thicket, with many thorns. To make a book of reasonable length and decent readability, I will take many examples from controversies and rulings on my local grid, ISO-NE.

I hope that, after reading this book, people will be more knowledgeable about the grid. And I hope that we ratepayers will understand how to take action on our behalf.

Even if we are not all insiders, we are all stakeholders in the grid.

CHAPTER 3

THE TWO GRIDS

Power and policy

THE GRID IS ABOUT KEEPING the lights on (I call this the Power Grid) and about payments and policies (I call this the Policy Grid.)

The first section of the book is about the Power Grid. I think of this section as being about reality and electrons. We need to know a little about the constraints of the Power Grid to understand the Policy Grid. However, most of the problems on the grid arise in the Policy section. Therefore, most of this book will be about Policy.

The power grid

THE POWER GRID IS ABOUT generators and voltage and wires. It is about delivering electricity to customers. This grid functions well, and when there is an outage, it is most often due to local weather conditions. Humans cannot change many of the constraints of the power grid. We can change our generation sources, but we can't change Maxwell's equations (electromagnetism) or Kirchhoff's laws (circuits). We can change fuels, but we can't change the laws of thermodynamics that govern getting "useful work" from heat engines.

These physical laws basically describe rules of nature that aren't going to change anytime soon. We will cover this topic only enough to see the constraints that physics puts on our power supply.

Be assured; this section will not contain equations or anything like that. Plus, it is short. I apologize in advance to my electrical-engineer friends for giving their expertise such short shrift in this book.

The policy grid: easy problems to hard problems

THE POLICY SECTION IS more complex. The RTOs have such overlapping thickets of regulation that it is hard to know where to begin.

I decided to start with a description of the energy auctions in RTO areas. We have to understand the way kilowatt-hours are bought and sold on an RTO grid before we can understand anything else.

After a brief introduction to the auctions, I will describe a comparatively simple problem on the grid and the complexities of resolving it in an RTO area.

Most people would agree that the grid should not cut people off from electricity in very cold weather. Electricity runs our home thermostats: that is enough reason to keep electricity flowing when it is twenty below outside. We will look at the RTO-induced complexities in handling a cold snap in New England.

The first thing that we will notice is that even an easy non-controversial problem becomes almost impossibly hard to solve in an RTO area.

At that point, we will plunge into the thicket of how the grid is managed and why it is going downhill in the RTO areas.

The grid we want

When I write that the grid is "going downhill," I need to be explicit about what kind of grid would be best. My three criteria are simple:

- Reliable electricity
- Relatively inexpensive electricity, so everyone can use electricity for their health and happiness
- Electricity made with low levels of pollution and low levels of ecosystem disruption

There is so much written about "microgrids" and "renewables on the grid" and "nuclear yes and nuclear no" that I feel it is important to state the goals that I hold for the grid. Not the methods (renewables, microgrids, demand response, nuclear, and so on) but the goals. When we move away from those goals, the grid is "going downhill." When we move away from these goals, the grid becomes more fragile, more expensive, and dirtier.

Again, I note: this book is not a complete and academic description of grid governance. People might fall asleep attempting to read a full-fledged description of grid governance in the RTO areas.

Perhaps people wouldn't fall asleep. If they truly understood what they were reading, they would become furious.

CHAPTER 4

THE ANGELIC MIRACLE OF THE GRID

Always in balance

REALITY AND ELECTRONS and physics. Let's start with the angelic miracle of the grid. "Electricity produced" and "electricity used" are always in balance on the grid. When you turn on a light, somewhere on the grid, a power plant makes more kWh for that light. When you turn the light off, somewhere on the grid, a power plant makes fewer kWh. Production and consumption are always in balance, in real time.

That is the angelic miracle of the grid.

Electricity is made in real time. Except for pumped storage and very limited use of batteries (more about them later), electricity is made, and it is used, simultaneously. Someone (a "balancing authority") has the responsibility for calling plants online and asking plants to leave the grid, in order to keep the grid in balance.

The grid is about power: the lights go on.

The grid is about policy: what sort of generation and who pays for it.

Keeping the power in constant balance is the angelic miracle of the power grid.

In contrast, some of the grid's complex policy schemes could justifiably be described as the work of the devil.

A fundamental issue that the grid must address can be described in the graph below showing the power required on the New England grid.[7]

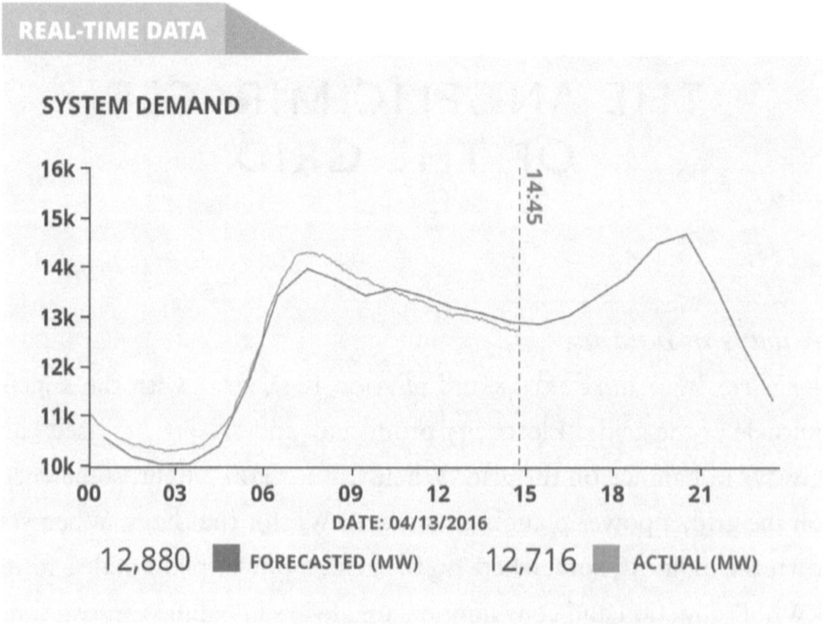

Figure 1: Electricity use on the New England Grid April 13, 2016 (ISO-NE)

People don't use the same amount of electricity every hour of every day. Looking at the graph, it is clear that more electricity is used at 9:00 a.m. than at 3:00 a.m. Therefore, more electricity needs to be produced at 9:00 a.m. To meet this varying need for electricity, some plants may run all the time; other plants will run only when called upon. The amount of electricity produced must be the same as the amount consumed. This must happen in real time. The grid must be in balance.

Let's look a little more closely at the graph. The numbers across the bottom are the time of day, expressed by a 24-hour clock. 3:00 a.m. is 3 o'clock, while 3:00 p.m. is 15 o'clock. The vertical axis is thousands of megawatts (MW) demand on the grid. One MW is a thousand kilowatts, and a kilowatt is a thousand watts. But we aren't done with the "thousands" yet. The vertical axis is in thousands of MW, (for example, 10k MW). A thousand MW is a gigawatt (GW). So, from watt to kilowatt is a thousand, from kilowatt to megawatt is another thousand, and from megawatt to gigawatt is another thousand. In general, watts and kilowatts describe various types of consumer use of electricity, megawatts are how power plants are usually described, and gigawatts are how the grid is usually described. In this case, for example, the number 12,000 MW (12 GW) is written as 12k MW on the vertical axis.

Now that we know the units, we can begin to describe the lines on the graph. The line covering the whole day is the demand ISO-NE predicted the night before (day-ahead market). The line stopping at 14:45 is the actual demand on the grid, as it evolved during the day. The vertical line is at 2:45 p.m., which is when I happened to take this screenshot from the ISO-NE website.

This graph describes electricity use on a mild spring day in New England. There are no particular demands or hardships on the grid. We can see that one-third more power is required at midday (15k GW) than is used all night (10k GW).

In some ways, the graph above is misleading: it shows 10K GW as the baseline. A more realistic visualization would show from 0 GW to 16 GW, instead of from 10GW to 16 GW. Such a graph would give a clearer view of how demand varies over the day. However, it would contain no more information than this one has. These graphs are used mainly for information, not visualization.

Another misleading part of the graph is that I chose a nice, mild day. On a day of high demand (midsummer with air conditioners or midwinter with supplementary electric heaters), the peak electricity requirement can be more than twice the nighttime requirement.

Could we even out this graph by using storage? If we could store electricity easily, many things would be easier. People are working on this problem: batteries, thermal storage underground, and so on. People have been working on storage since I entered the workforce, more than forty years ago. Some improvements have been made, but the barriers are huge. Storing electricity means losing some energy on the round trip: power into the storage, power out of the storage. This will waste some power. Some power absolutely will be lost to the round trip. Also, if we choose battery storage, we have to realize that manufacturing batteries is resource intensive. For example, it would take vast amounts of specialty mining if we decided that we needed to build lithium batteries at grid scale.

A decision to store power is an engineering and economic decision. Is investing in resources for storage and losing power on the storage round trip actually a better deal than making power as needed, as is done now? A "better deal" does not necessarily just mean a better *economic* deal. If a low-carbon grid, but without nuclear energy, were to become one of the criteria for grid operation, then some kind of energy storage could be considered part of a "better deal." However, people who are sure-in-their-hearts that renewables can do everything usually assume that the storage problem is solved somehow. It isn't. And such people rarely acknowledge that any storage solution will be subject to the materials-availability issue and the round-trip-power-loss issue. Moreover, to store electricity and use it later, we will have to make more electricity than if we used it immediately.

In this book about the grid, I am going to talk about real time and real systems. In other words, since right now grid-level storage is not an option, I will not write as if it were available.

Transmission and distribution

So far, I have been talking about electricity generation. To get electricity to your house requires more than generation.

Most of this book is about electricity in areas with Regional Transmission Organizations (RTOs). Before RTOs were created, however, utilities were vertically integrated, and the same company that owned the power plants owned the substations and distribution system. The distribution system consists of the electricity lines and small substations that bring the power to your house and neighborhood. In contrast, a transmission system is a bigger system. The iconic image of a transmission system is a set of huge towers and transmission wires, usually starting at a power plant and extending for miles across the country.

Transmission lines extend over great distances. Even before RTOs were invented (in the late '90s), it would have been rare for a utility to own all the transmission lines in an area. Also, transmission lines often cross state boundaries, which presented other governance problems.

In terms of transmission lines, especially, utilities co-operated through regional power pools. This was a relatively simple concept, though it definitely had problems.

If you were a new kid on the block, a new company building a power plant, nobody guaranteed that you could break into the transmission system. After 1973, however, if you built a power plant, you could break into the system. This was due to a Supreme Court ruling against Otter Tail Power.[8]

Otter Tail had refused to carry power (wheel power) from a Bureau of Land Management hydro plant to the territory of a municipal utility.

The two facilities were in neighboring states, which is probably one of the reasons that this became a Supreme Court decision. Since Otter Tail refused to use its transmission lines to transfer power from the hydro plant to the municipal-utility territory, the municipality had to buy power from Otter Tail itself. (In the utility business, this type of refusal is called "refusing to wheel power.") Otter Tail owned the transmission lines and decided who could use those lines.

In the court case, the Supreme Court ruled that the transmission lines were an "essential facility" for business. Otter Tail could not refuse to wheel power for the municipal utilities. The decision was lengthy and somewhat confusing. Supreme Court justices agreed in part and dissented in part. Some parts of the ruling were based on the Sherman Anti-Trust laws, and some of the dissents were about the relevance of the Sherman Anti-Trust laws.[9]

I was active in renewable energy in the late '70s and early '80s. Looking back at the '80s, power companies were concerned that their transmission lines would become "common carriers" and endure federal regulations, like other common carriers such as trucks. The Otter Tail decision stopped just short of the "common carrier" designation for transmission lines, so everybody was sort-of happy. There have been many federal rulings and decisions since Otter Tail, and the decision is no longer relevant on most grids. Wheeling power has been facilitated by several later decisions.

Most of this book will be highly critical of RTO governance, but I wanted to write a little about Otter Tail. I needed to show that vertically integrated utilities were not a perfect situation. No, everyone did not just get along wonderfully back in the old days. However, vertically integrated utilities did have some important advantages over the RTO system.

The main advantage of the vertically integrated systems is that those systems had clear lines of responsibility.

Now that we are talking about transmission ...

In general, transmission systems follow the power plants, and so the closing of power plants and the building of "distributed generation" (smaller power plants, such as solar arrays) increases the cost of transmission. Renewables advocates will disagree with this statement. Indeed, if renewables were always available to be dispatched to the grid, this statement would be incorrect. However, since most renewables are available only part of the time, they must be backed up with other forms of power. Therefore, transmission costs *do* increase. There will be more about this in later sections.

To move power long distances, with relatively low line loss, you need high voltages. That is why you can see really tall transmission towers crossing the countryside. The height of the tower is a rough indication of the voltage of the lines. Taller towers mean higher voltages and, therefore, less line loss. Nobody would build such high towers unless they were moving a lot of power.

What is line loss? As you move power, some of it is wasted as it moves. In general, power is wasted two ways: resistance and electromagnetic radiation. Resistance heats up the power lines. Electromagnetic radiation (radio waves) are the "static" you hear when your car drives under a power line. Both these phenomena are well understood by electrical engineers and cause very few problems for the grid. However, line loss does make moving power more expensive, since not all the power will arrive at its destination. As a rough rule, the longer the line, the more power will be lost.

As is common in most of the issues about the grid, the movement of power is well understood and simple, while the movement of money (policy) gets more complex and arcane every year, especially in the RTO areas.

A friend of mine, familiar with both phone and electrical deregulation, summed up deregulation this way. Electricity generation is a simple technology with a complex set of rules, while the phone system is a complex technology with a simple set of rules.

CHAPTER 5

THE BALANCING AUTHORITY

Electricity must be made in real time

THE REQUIREMENTS FOR electricity on the grid are neither constant nor fully predictable, and electricity must be manufactured and then used within milliseconds. Making more or less electricity than is immediately used will mess up the grid, first by changing the frequency of the electricity, and later by causing other problems, such as area-wide blackouts.

Actually, milliseconds between production and use is too long a measure, because electrons in wires "travel" at a speed close to the speed of light. However, it can be misleading to say that electricity must be used "instantaneously" or "immediately." Most people consider all sorts of things to be "instantaneous" when, in reality, they take seconds or minutes or even longer. "His statement had an instantaneous effect on the crowd."

This problem with the word "instantaneous" reminds me of the standard anti-nuclear statement that humankind has never encountered a poison like "radioactivity that will be dangerous for thousands of years." Indeed, humans are more familiar with poisons that will be dangerous forever, not just thousands of years. Think arsenic: it doesn't become less poisonous with time. Arsenic is poisonous forever. However, for most people, "forever" seems shorter than "thousands of years." Similarly, "instantaneous" seems longer than "milliseconds."

Well, enough about words and time. Back to the balancing authority.

The work of the Balancing Authority

ALL GRIDS HAVE A Balancing Authority (BA). The BA organization is often part of a bigger multi-state organization, such as the New England grid operator, ISO-NE.

The BA makes sure that the supply of power on the grid is exactly matched with the requirement for power. Always. As we can see in figure 1, the grid's requirements for power varies during the day. Power plants must be called online as the requirement rises or asked to stop producing electricity as the requirements fall.

But that is not the BA's only problem. Another problem is that different technologies have different abilities. A steam-fired power plant can put out more or less power but must change power levels relatively slowly. It runs wonderfully when it runs steadily, but it takes some time to turn around. By design, a steam plant is a steady, solid producer. Of course, a steam plant can change the output. We wouldn't be able to have nuclear submarines (full speed ahead!) if steam plants couldn't change their outputs.

Still, an internal combustion engine is far more nimble than a steam plant. The internal combustion engine in your garage is probably capable of zero to sixty in less than thirty seconds. A recent-vintage

Camaro is capable of zero to sixty in four seconds, and proud of it. No steam-powered plant is likely to be able to move so quickly. I think nuclear subs are probably the fastest at changing power levels, but I also think the speed at which they can change power levels is classified information.

Meanwhile, on the civilian grid, steam cycle plants change output slowly, while internal combustion plants (gas turbines, diesels) and hydro plants can ramp up and down quickly. The Balancing Authority operator has to take into account the fact that demand changes, and some power plants are quicker to ramp up or ramp down than others.

But wait, there's more. (There's always more.)

Even the internal combustion plants are not "instant" the way you need to have electricity be "instant." So there's this whole other thing called "ancillary services" on the grid: this basically consists of paying plants to be on various types of hot standby, often with turbines spinning (but no load), ready to send their power to the grid very fast, when called on. Of course, plants can't just keep their turbines spinning out of the goodness of their hearts: they will have to pay for fuel. The Balancing Authority has to arrange for spinning reserve and pay for it. Spinning reserve, which is fast to get on the grid, will cost more than slightly slower-to-the-grid spinning reserve. How much spinning reserve does the Balancing Authority need at any particular time? What kind of spinning reserve? As they say on Facebook: it's complicated.

So far, I have written about conventional power plants on the grid: steam turbines, diesels, and hydro plants. What about renewables?

Renewables often make the BA's job harder. They depend on real-time conditions (sun, wind), and the BA cannot order them online to match requirements. The BA can ask them to get offline (called "curtailing"), however. So the BA's options are more limited for solar and for wind. As a general rule (there are some exceptions), a BA cannot control *when* customers use power. With wind and solar, the BA also cannot control when power is available.

There are other problems. Wind turbines are often placed where there are not enough transmission lines to carry the power from the turbines to the load centers, especially in periods of high energy use. Wind is also famously spiky: while grid demand changes slowly, the wind starts up and dies down with comparative suddenness. Poor transmission connections plus spikiness mean that the wind is frequently curtailed. The wind may blow hard, but the local lines are full. At that point, the grid operator asks the wind operators to partially or fully disconnect from the grid (curtails the wind installation). This sort of request is a standard responsibility for the BA. The grid must be in balance.

With solar, the course of the sun across the sky can form a "duck curve" on the grid. The duck curve happens on a sunny afternoon when demand on the grid is low because so many solar installations are providing power to homes and to the grid. In the afternoon, solar contributes power to the grid. But eventually the sun goes down, and thermal plants have to ramp up quickly to provide the power that solar had provided.[10] This ramping requirement is called the "neck" of the duck curve.

The neck of the duck curve is becoming an issue in areas with a lot of solar. Solar input to the grid tends to be highest during summer afternoons. However, when the sun goes down, the solar goes offline rapidly. The BA then orders the dispatchable plants (thermal plants, hydro plants) to ramp up, and they often have to ramp up faster than the solar is ramping down. Faster because people have a tendency to turn on a light as the sun sets, or come inside and begin cooking dinner, and so on.

There's a rule of thumb on the grid that no plant should be so big that it is more than 10% of the average demand on the grid. One

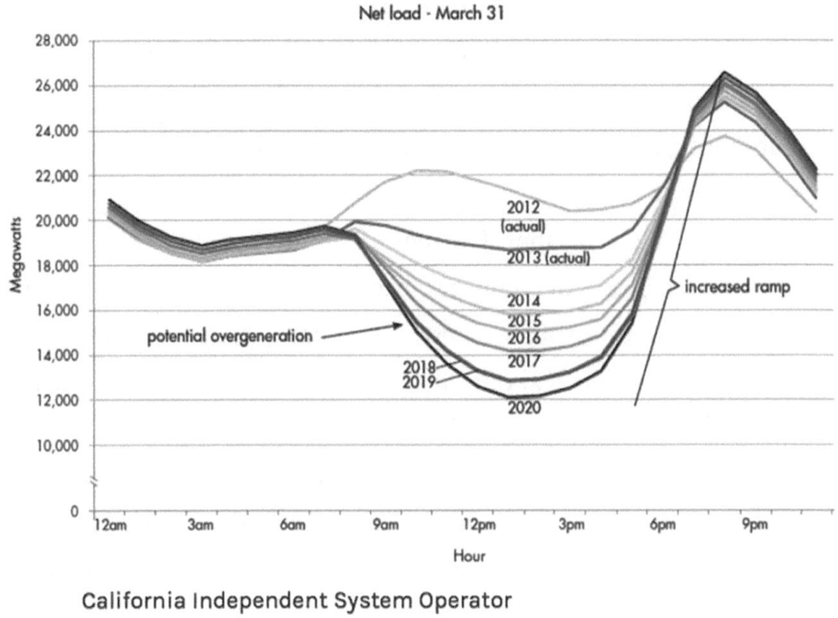

Figure 2: Duck curve on the California grid (Department of Energy)

plant going offline should not take 20% of the grid's power with it. People look at solar as a distributed system: my rooftop, your rooftop, and a solar array down by the Interstate. No huge power plants here! However, in fact, solar often acts like a single megaplant, which switches off in the early evening. This is another issue that the BA must face.

And finally, we get into Volt-Ampere Reactives (VARs). We are all familiar with voltage: for example, a high-voltage line or a 12-volt battery. We are mostly familiar with amperes, a measurement of current. For example, a circuit-breaker in your home will trip if one of your circuits is attempting to carry too many amps. But there is another property of electric current, a property with which most of us are not familiar: the VAR. The BA must keep voltage within a narrow range and balance demand on the grid (amps). The BA must also make sure the VARs are in balance.

Short, geeky interlude

Alternating current has electromagnetic properties that have to be kept in balance. When a person attempts to explain VARs, they often end up giving incomprehensible explanations, full of sine waves and imaginary numbers.

I'm not going to do that. Citizens can understand the electric grid without being electrical engineers, because most of the grid problems we face are policy problems, not electrical engineering problems. However, this section is about the physical grid and the job of the BA. Therefore, I need to mention those VARs.

I found a wonderful analogy for VARs in a tweet stream by @ElephantEating. The person behind ElephantEating is Eric Hittinger, a professor at Rochester Institute of Technology and an avid bicyclist.

To explain VARS, Hittinger makes an analogy to riding a bicycle. The energy you put into the pedals will move the bike forward, but you also have to put some energy into maintaining your balance, or you'll fall over and won't be able to move forward at all. If you are a good bicyclist on a smooth road, the "maintaining your balance energy" will be small. If you are a poor bicyclist who swerves around a lot, or if you're on a bad road, the "maintaining your balance" energy will be larger. In either case, the "maintaining your balance" energy is necessary. That energy is also a parasitical drain on your energy effort: it doesn't move the bike forward.

A well-run grid is like a good bicyclist on a smooth road, while a more-difficult grid (more sudden ups and downs in power or energy requirements) requires more of the balancing-type energy. Fabulous analogy!

The BA is also responsible for balancing VARs.

End geeky interlude

Rotating electric machinery puts VARs on the grid, and if the entire grid was thermal (nuclear, gas, coal) and hydro units, there would rarely be a problem with VARs. These systems all run with rotating electric machinery.

However, wind turbines and solar make direct current that needs to be changed into alternating current, and that process does not put VARs on the grid in the same fashion. (Some older and bigger wind turbines do put VARs on the grid.) At any rate, messing up the VARs can also mess up the grid, so this is another place where the BA must be aware of what is happening on the grid.

Vermont had an interesting illustration of this issue. The Kingdom Community Wind Project connected to the grid without installing synchronous condensers. The developer said that it would install such condensers, but the wind project connected to the grid without them. Without the condensers, the grid operator had to curtail (not accept) the wind power quite often, mostly due to VAR mismatch. This curtailment of wind led to an acerbic exchange between the grid operator and then-governor Peter Shumlin.[11]

Shumlin was angry (and a bit accusatory) because the BA had curtailed the wind turbines and, therefore, had curtailed a Vermont renewable-energy installation. Didn't the BA know that Vermont needs all the renewable energy it can obtain? Shumlin wrote a cranky letter to the BA, and the BA wrote right back.

However, this exchange of letters meant very little. Even if a governor is annoyed, the BA has to do what they have to do to keep the grid in balance. And they will do it.

A sort-of happy ending to the Kingdom Community Wind Project story: the owners invested $10 million in synchronous condensers, the VAR problem was mostly solved, and the wind turbines were not curtailed as often after the condensers were installed.[12]

The physical operation of the grid faces more challenges than I have listed here, including gas-fired plants not getting fuel supplies when home heating is using lots of gas on cold winter days, and so forth. However, in general, physical issues are well understood by engineers and operators. The BAs know how to keep the lights on, and they generally do it.

However, policy and payment issues are growing, and these issues are marching over into the management of the physical grid. If it were just a matter of power plants and engineering, that would be one thing. But it isn't. So, we must reluctantly leave the optimistic area of happily balancing physical constraints. The moment-by-moment successful balance on the grid makes every functioning grid into one of the wonders of the modern world.

We must now move to the more depressing area of payment and policy, which is quite capable of wrecking what the BAs and engineers have achieved.

CHAPTER 6

VERTICALLY INTEGRATED OR RTO

Power and payment

ELECTRICAL ENGINEERS WHO specialize in the grid learn how to handle many complex issues: balancing users with suppliers, keeping the "imaginary" part of the grid (VARs) in balance while keeping the visible part of the grid (voltage and frequency) in balance. The grid is a modern miracle. It takes technical knowledge, computer power, and skill to operate it.

In contrast, in many parts of the country, understanding how the power is paid for is a black hole of impossible acronyms and "I can't believe this" payments. These parts of the country are the Regional Transmission Organizations (RTO) areas. Understanding the RTO payment system requires insider knowledge, willingness to use acronyms without examining them too carefully, and a cheerful willingness to spread high prices to low-income parts of society.

What are the RTO areas? How did we get RTO areas?

The answer to the first question is easy. Here's a map of the North American RTO areas. The RTO areas are the shaded sections of the map.[13]

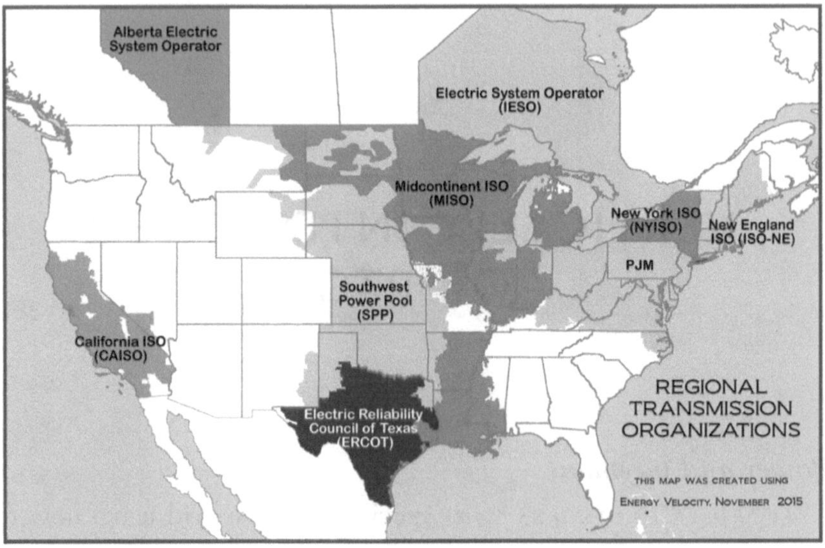

Figure 3: RTO areas of North America (FERC)

RTO areas began with FERC order 2000, issued in 1999, which allowed RTOs to form.[14] Clearly, not all areas of the country chose to join RTOs.

Confusingly, some RTOs are called Independent System Operators (ISOs). In this book, we will refer to all such organizations as RTOs, though, of course, when a specific name contains "ISO," we will use that name.

Before the RTOs

LET'S START BACK IN THE days before there were any RTO areas. Before 1999, all utilities were "vertically integrated." In much of the country, this is still the case, as we can see by the non-shaded sections in the map above.

Before 1999, there were only vertically integrated, utility-owned power plants and distribution equipment (local wires and substations). The utility usually either owned or had cooperative agreements about transmission lines. The utility had requirements to keep the lights on: it reported to a local board of some kind, and it would be judged and possibly fined if there were too many outages.

The utility also got paid a rate-of-return on its expenditures. For example, if the utility had to build a power plant, it was allowed to charge enough money to build it and to make a rate-of-return profit on building the plant and running it. However, in many jurisdictions, a utility's rate-of-return could be lowered if its customers' lights went out too often.

So, how would a utility attempt to make more money? First, let's look at that "rate of return." The regulatory agency might choose to lower the utility's rate of return if the customers suffered too many power outages. This was often done by fining the utility for poor reliability.

Therefore, before RTOs, it was in the utility's best interest to spend "whatever it takes" to keep all the power plants and transmission systems in top shape and to have crews ready to fix problems very quickly. In other words, the higher the reliability, the better the chance of a high rate of return on the utility's investment.

The second way for a utility to make more money would be to invest more money, resulting in a bigger "rate base," as it is called. It was in a utility's best interest to build more power plants, run a visitor center, replace worn-out poles, and so on. It all went into the "rate base," and the "rate of return" percentage calculation was based on the rate base.

Someone noticed (many people noticed) that this was basically a perverse system. The way for a vertically integrated utility to make more money was ... to spend more money. Unlike Joe's deli down the street, which was always trying to be more efficient with money, in order to have larger profits, a vertically integrated utility wanted to be less efficient with money, so they could have a bigger rate base and larger profits.

One of the ways that utilities made more money was to overbuild power plants, that is, build too many power plants. Presenting a somewhat inflated projection of electricity needs in the future could justify a new power plant, which would be approved by the local regulators. The cost of building the plant would go into the utility's rate base. In this sense, a new power plant generated utility profits, whether or not the electricity was needed. Utilities always need more power plants than their peak power use (they have a reserve margin of plants), but under vertical integration, the reserve margin tended to grow along with the rate base and the utility profits.

Now, of course, the utility couldn't just build things and spend money. A state regulatory agency had to approve the expenditure. Depending on the state, this agency was usually called the Public Utilities Commission (PUC) or the Public Service Board (PSB). At the state level, utilities tended to get a little cozy with their regulators ("regulatory capture"). Overbuilding and overcharging customers was often the result. On the other hand, the utility had incentives for building a very reliable grid.

I was a project manager at the Electric Power Research Institute (EPRI) when utilities were vertically integrated. I remember how we went to all-hands meetings. At these meetings, we were exhorted, "When referring to people who purchase electricity, call them 'customers,' not 'ratepayers.'"

But *everyone* called the customers "ratepayers." We knew that the utility and its PUC decided on the electric rates and that the ratepayers dutifully paid them.

CHAPTER 7

RTOs REPLACE THE INTEGRATED UTILITIES

Market rides to the rescue?

LOOKING AT THE PERVERSE incentives for integrated utilities, many people had the idea that a market would fix these high prices—if the utilities could just be forced to compete in a market. Joe's deli can't raise lunch prices without losing customers. The hope was to set up market constraints for utilities, similar to what Joe faces in running his deli. The idea was that, if utilities participated in a market economy, resources would be used efficiently, and customers would be served better.

But even then, even in the 1980s, when utilities were vertically integrated, the dark side of "saving money" was apparent. Some academic studies and EPRI studies implied that the grid was over-designed, overly expensive, and unreasonably reliable.

I found some of the "overly expensive and unreasonably reliable" talk appalling. I didn't like the idea of saving money by lowering reliability.

RTOs: Yes, a market? No, not a market?

Remember the ratepayers? You know, the consumers of electricity? One would think that deregulating the utility markets would mean the consumers are no longer ratepayers. Those ratepayers, captive to their local utility, would actually become customers—customers who made choices.

Customer choice is one of the most important ways that markets work. That is how deregulation of airlines worked: choose your flight, and airlines can now compete on price. That is how deregulation of the landline telephone companies worked: choose your long-distance carrier, and different carriers will offer different programs.

That's not how utility deregulation turned out. Deregulation did not give consumers a choice. In most cases, they remained the same old ratepayers, captive to their providers.

The American Coalition of Competitive Energy Suppliers (ACCES) has compiled a state-by-state listing of energy choices.[15] Only eighteen states (including the District of Columbia in this count) have electricity choices. In most of these states, only *some* consumers are allowed to have a choice. The consumers who have a choice may be large industrial or commercial consumers, or all consumers in a utility's service area.

Different states have different rules, and the rules do not correlate with the RTO and non-RTO areas. For example, Connecticut and Massachusetts have electricity choice for residential consumers, but only in the territory of *some* utilities. Vermont has no electricity choice for consumers, although it is in the same RTO. In California, electricity choice is available only for a capped percentage of commercial and industrial customers, in the territory of *some* of the distribution utilities. In other words, in California, *some* big customers get *some* choices.

It gets more complicated. A careful comparison of the ACCES listing and the RTO areas shown in figure 3 illuminates the complexities. Indiana is shown as having a choice in the ACCES list. Indiana lies in MISO territory, but consumers have a choice of only natural gas providers, not electricity providers. Meanwhile, Illinois and Ohio, also in MISO, border Indiana. ACCES lists those states as having choices for electricity providers.

Only some of the states in RTO areas have electricity choices for all consumers. Most ratepayers in RTO areas do not have a choice of utilities. Ratepayer choice is an add-on, a maybe-nice-to-have, not the central feature of RTOs. In short, utility deregulation was not about the ratepayers. It wasn't about turning ratepayers into customers who made choices. So why would we expect it to be good for the customers?

Spoiler alert: Utility "deregulation" hasn't been good for the customers. As a matter of fact, it isn't deregulation. The regulations for RTO areas are longer and more complex than those for areas that didn't choose to form an RTO. Also the RTO areas have higher prices for electricity than non-RTO areas. Therefore, calling the RTO areas "deregulated" is a misnomer. In this book, I will call them "RTO areas."

Another question needs to be answered: Assuming RTO areas are some kind of market, would this market keep reliability high? Airline deregulation kept safety, and phone deregulation kept reliability. Would RTO areas keep reliability?

Mostly, reliability remained high. In an RTO area, there are basically three types of utilities.

- Generation utilities own facilities that generate power. They are often called "merchant generators."
- Distribution utilities own the distribution systems that distribute power.
- Other types of load-serving entities (LSEs).

Some RTOs have LSEs of various types—companies that buy energy and resell it to customers. Such companies are available in only a few states and are not a central feature of the RTO areas. Many LSEs do not maintain distribution systems. Instead, they use the distribution systems of the "regulated" distributors, but they make their own contracts with generators.

Distribution utilities are usually "regulated" in the same way as vertically integrated utilities have been "regulated": with a rate of return. In contrast, generation utilities are supposed to compete in energy markets and auctions. Independent LSEs provide customer choice, but they are available in only some parts of some states.

The central features of RTO areas are merchant generators, which are separate from distribution utilities, and generators that compete through auctions. Customer choice is sometimes added to the mix, but it is not central. In this book, we will describe the more universal features of the RTOs (auctions). Whether or not states have some customer choice within an RTO, ratepayers pay more in RTO areas, as described in chapter 13, "The Death of the Market Hope."

So, what about reliability in RTO areas? It has been a mixed bag, for sure.

There was a long history of overbuilding power plants before the RTO areas were founded, so lack of power on the grid was not a problem in the early years.

The distribution system causes the most frequent types of power outages. Wind, ice, and tree branches disrupt utility lines. In the RTO areas, the distribution utilities are regulated, and they have the same incentives for reliability that they had before. They get a rate of return on their expenses. As expected, most distribution utilities continue to do a good job. In some areas, distribution utilities also own power plants, which compete on price as merchant generators. In these areas, the distribution system can suffer under deregulation. If the generator portion of the company is losing money, distribution is sometimes an easy place to cut expenses.

With RTO areas, merchant generators can game the system and cause power outages. When this happened in California, it caused rolling blackouts. But that is getting ahead of ourselves. Right now, we will look at ordinary operations in the RTO world.

In RTO areas, distribution utilities buy power from generation utilities. The distribution utilities are still regulated monopolies, with every incentive to keep their power lines and substations in good repair.

The distribution utilities buy the power at auctions run by the RTO. In these auctions, the power plants compete with each other in auctions that run at five-minute intervals. The "queue" (dispatched plants) is the set of plants that bid the lowest price per kWh, and the plants in the queue provide enough power for the current real-time electricity requirements on the grid. If the load requirements are low and a plant's electricity is expensive, the plant won't make it into the queue at that time. The power plants want to make it into the queue. They want their power to be dispatched, and they want to be paid.

Actually, it is more complicated than that, but for now, understanding the RTO auctions is the main thing. The distribution utilities buy kWh from the generators, and the RTOs run the auctions.

With the auctions, power plants face a new situation, and the plants have very different incentives than they had within vertically integrated utilities. The auctions between power plants removed each power plant's incentives for reliability. As a matter of fact, the grid itself can become less reliable. In my area, studies of the future of the grid show the strong possibility of rolling blackouts in winter weather, due to insufficient generation to meet the high demand. In other words, the RTOs can enable a new type of reliability problem.

RTOs have not been good for the consumer. They gave us higher prices and the possibility of a new kind of reliability failure.

In RTO areas, as in vertically integrated areas, very few of us have choices for where to buy our power. We are not yet consumers. In the RTO areas, we are still ratepayers.

RTOs and reality

MUCH OF THIS BOOK will be about grid governance in RTO areas. At this point, however, we need to start with an example. Even relatively straightforward grid management gets difficult or impossible in an RTO area.

The example is about winter in New England. It's important to keep the lights on during winter cold snaps in the northern parts of the country. Electricity runs home thermostats and fans and pumps. It is not reasonable to expect everyone in New England to heat with an old-fashioned wood stove, complete with a woodpile out in the woodshed. New England isn't just bucolic farms: it's also Boston and Hartford and Providence and New Haven. For most people in New England, being safely warm in the winter depends on electricity. In New England, in winter, a fragile, unreliable grid is dangerous.

Therefore, for an example, and for a reality check on the RTO areas, we will watch ISO New England try to keep the lights on in New England during a cold snap.

CHAPTER 8

WINTER LIGHTS

Keeping the lights on in winter

MOST PEOPLE WOULD agree that, during winter in the Northeast, grid operators should not disconnect customers from the grid. In cold climates, being disconnected in winter is dangerous. Losing electricity in winter usually means that the house becomes cold. Without electricity, thermostats won't work, and fans and pumps won't work (traditional heat sources such as forced air and hot water heating). More modern methods won't work, either. Pellet stoves and cold-weather heat pumps need electricity. A traditional wood stove will work, but "everyone needs to have wood stove heating" is not the solution for an entire region or a country.

I have friends who have their own generators for backup. However, asking everyone to buy a diesel generator is not a civilized way to run a grid. Generators are expensive, and they need to be kept in good trim.

They are not suitable for everyone. They are completely impractical for people who live in apartment buildings.

Also, though a generator can make a short outage more bearable, a generator is only a short-term fix. When the grid is down, there will be no electric pump working at the gas station to refill your diesel. If the grid is down, refilling your home generator will be a major problem. In other words, even with a home generator, after you have used the fuel you store at home, you are still dependent on the grid.

We need a reliable and steady supply of electricity in cold winter climates. Now let's look at the powerful problems that cold weather presents and how the RTO attempts to deal with them.

Gas just in time

OUR NORTHEASTERN GRID is heavily dependent on gas-fired power plants, which make more than 50% of the electricity on our grid. The percentage of electricity from gas-fired generators has steadily risen over the years. Some of the reasons are simple. Gas has become less expensive (shale revolution), and coal and nuclear plants are being retired. The retirements are happening for various reasons, but one reason is the availability of inexpensive gas. Another reason is that, in the RTO areas, capacity payments are structured in a way that they provide a high level of support for gas-fired plants. For now, however, we will concentrate on the facts (a grid that is overly dependent on natural gas) rather than the causes.

For the grid operator, this dependence on natural gas is a potential *and* actual problem. Gas is delivered through pipelines: it is just-in-time delivery. There is very little storage for gas at a power plant, though there is storage to feed the pipelines. But no matter how much storage can feed the pipelines, the pipelines can carry only a certain amount of gas at a given time.

Since electricity is made and used instantaneously, and natural gas for power plants is delivered just in time for use, sometimes the two "just in times" don't mesh, and that leads to trouble.

New England winter and gas use

Power plants receive natural gas through pipelines, a just-in-time delivery scenario. In New England, in summer, the pipelines can deliver enough gas for the power plants, even on a hot day with high electricity demand on the grid. However, the pipelines cannot deliver enough gas to the power plants in winter, because many homes are heated with gas. Homes have first priority for natural-gas supplies. On a cold day, the pipelines cannot meet both sets of needs: they cannot deliver enough gas to homes for heating needs *and* simultaneously deliver enough gas to power plants to make electricity. In this situation, gas-fired power plants may not be able to get fuel.

Next, we will see the reality of what happened in a recent cold snap in New England. Then we will see how the RTO responded to that reality. We'll start with an op-ed piece I wrote for our local paper, just after a major cold snap.

CHAPTER 9

OIL SAVES THE GRID

Around 5:00 p.m. on January 6, 2018 I snapped a light on as the sun went down. The temperature was around minus 8 degrees Fahrenheit. It had been zero at lunchtime and would be minus 15 the next morning.

As usual, the light went on. As grid operator ISO New England had planned, oil had saved the grid.

During that very cold week, about one-third of New England's electricity came from burning oil. The people at ISO-NE might think it is unfair to say that they planned to save the grid with oil, but they did, because of the Winter Reliability Program. But first, some background.

Over the past few years, as the price of natural gas fell lower and lower, more and more natural gas has been used on the New England grid. Gas-fired plants could undercut the bids of other plants. Therefore, gas-fired plants ran more hours, and new gas plants were built. Though the overall demand on the New England grid has been virtually flat for years, more and more gas has been used to make electricity. In 2016 (the last year for which full data is available), 49% of our electricity was fueled by natural gas.

In many ways, having so high a percentage of natural gas on the grid was "cruising for a bruising." Electricity must be produced at the exact moment it is needed. Oil, coal, and nuclear plants are prepared for demand variations because they store fuel on-site. In contrast, gas plants are supplied by pipelines: They get the fuel when they need it — if the pipeline has it available. Using natural gas for electricity means that real-time electricity meets real-time pipelines. Everything has to happen *right now*. In some circumstances, it was clear that this would not end well. Homes use natural gas for heat, and homes have priority on the pipeline. In very cold weather, when homes need a lot of gas, power-plant gas supplies are interrupted.

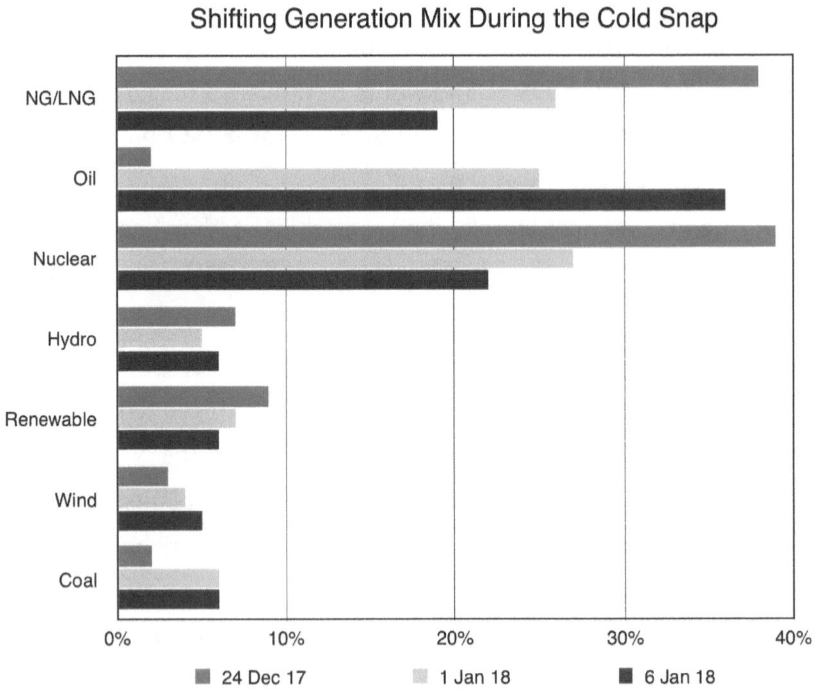

Figure 4: Fuel mix moves to oil during the cold snap (ISO-NE data)

The Winter Reliability Program exists due to the foresight of the grid operator, and this program pays for oil. Many gas-fired plants can also burn oil if they have it, and the bulk of reliability-program money goes toward paying such gas-fired generators to keep oil on-site. In the early evening of January 6, 2018, when I switched on the light, the grid was running about 30% oil-fired. It had been running at that level of oil consumption for several days. The figure on the previous page (New England fuel mix moves to oil) illustrates how much oil was used on the grid, as gas became more expensive and less available.[16]

Side note

The national grid regulator, the Federal Energy Regulatory Commission, does not like the New England Winter Reliability Program because it does not like grid operators to pay directly for one type of fuel. Next year, there will be no such program. Instead, ISO-NE will implement an untested program of fines and rewards for power plants (known as Pay for Performance), aimed at keeping the plants online in bad weather. Since the grid itself will not change, I think the new program is unlikely to change the amount of oil used on the grid.

And a note about price

In the cold weather, the price for gas was soaring. The usual daily price for natural gas in this region is about $4 per MMBtu. On January 4, 2018 gas was $87 MMBtu. (MMBtu stands for 1 million British thermal units. One Btu is the amount of heat required to increase the temperature of one pound of water 1 degree Fahrenheit.)

Electricity prices on the grid also ran very high. The usual grid prices of 2 cents to 8 cents per kilowatt-hour went up to 15 cents to 40 cents and stayed high for days. (A kilowatt-hour, or kWh, is the amount of electrical energy consumed when 1,000 watts are used for one hour.)

Expect all this to be reflected in your electricity bill, by the way.

End of note about price

Back to our grid, which was more than 30% oil-fired. All that oil made our winter grid more carbon-dioxide heavy than our summer grid. In planning for the future, I believe we should also think about the carbon content of the fuel. How were our low-carbon sources of energy doing in the cold weather?

Nuclear plants made about 20% percent of the New England grid's power, despite the fact that the smallest plant (the Pilgrim Nuclear Power Station in Massachusetts) went offline after one of its power lines went down.

Wind turbines did very well while the wind was blowing. During those times, the grid received about 5% of its power from wind turbines. Unfortunately, a lot of the time, the grid received less than 1% from wind.

Hydro, biomass, and refuse burning added up to about 10% of the power, and they were very steady.

It is very hard to tell how much solar is on the grid, because much solar is "behind the meter" and not tracked well. I would say, however, that with snow on the solar panels, we probably had less solar than usual. So, depending on the wind, we had about 35% of our power from very low carbon sources. And yet, there was all that oil...

What should we do to avoid having a high-carbon grid in bad weather? The answer is simple but not easy to implement. Basically, the way to have more low-carbon sources on the grid in bad weather is to have more low-carbon sources on the grid in good weather. Plants get stressed in various ways in bad weather, and all sorts of plants go offline. To have a low-carbon winter, we need low-carbon summers.

This will not be easy to do. We can start by not closing existing low-carbon sources (nuclear plants, hydro, wind turbines) and look at building more. However, low-carbon sources are not locally popular, and most people don't even know about the oil-burning on the grid in bad weather. So, I am not hopeful. I think there is more oil on its way for the New England grid.

I see some hope, however.

Other northern areas have low-carbon electricity, with almost no fossil fuels. Both Sweden and Ontario have very low-carbon grids, based on nuclear and hydro. Sweden and Ontario emit less than half the amount of CO_2 per kWh as New England or Germany. These areas are also adding wind, but the basis of their low-carbon grid is nuclear and hydro.

I wish the Northeast would follow the example of these low-carbon areas, but I fear that we will not.

****

I wrote the material above as an op-ed for my local newspaper, the *Valley News*. It was printed on January 27, 2018.[17]

Oil almost fails at saving the grid

DURING THE COLD SNAP, things were worse than I had reported. On January 12, ISO-NE wrote a report on their Cold Weather Operations between December 24, 2017, and January 8, 2018.[18] The ISO-NE report included several areas I had not covered. (I had not realized that ISO-NE would be issuing a report a few days after I wrote my op-ed.) This report showed the fragility of the grid much more clearly than my op-ed had shown it.

As I noted in my op-ed, the grid operator had put a "winter reliability program" in place to pay for oil for dual-fueled generators. The Winter Reliability Program paid for oil, and oil saved the grid. I knew that, but I didn't realize that oil inventories at power plants were sinking fast. Even with the reliability program, dual-fuel generators had not stored enough oil for the situation and were having trouble trying to refill their oil tanks. The ISO report shows the usable oil inventories at power plants sinking steadily from around four million barrels available on December 1 to around one million barrels on January 8.[19] Luckily, the weather warmed up at that point.

Figure 5 shows the oil inventory at a specific station. The inventory fell from a seven-day supply on December 1 to slightly more than a one-day supply on January 8.

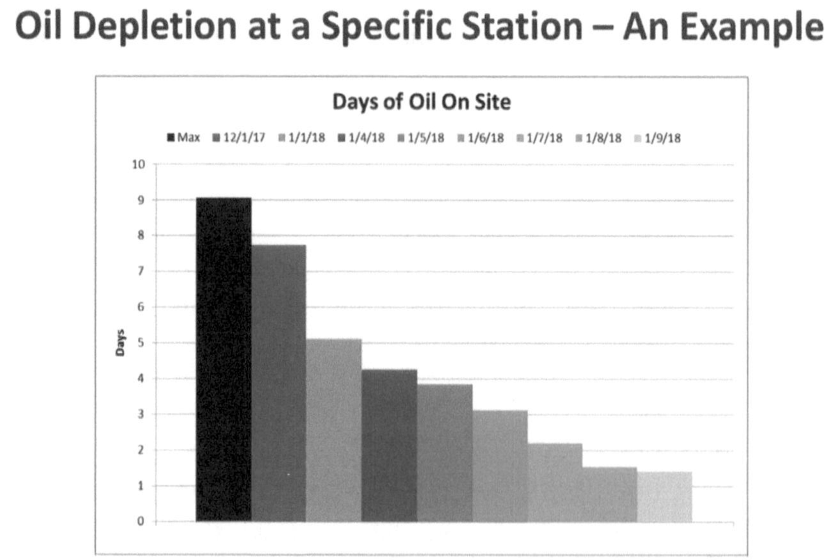

Figure 5: Days of oil stored at a power plant as cold snap progresses (ISO-NE)

In my op-ed about the grid in cold weather, I wrote that ISO-NE had the foresight to pay for oil storage at gas generators that could also burn oil. Did this matter? Would those generators have stored oil on-site without ISO-NE support?

Well, no. They wouldn't have stored oil. This is partially speculative: we can't do a complete comparison of "what happened" with "what might have happened under other circumstances." However, some facts do jump out at us.

The plants that were in the Winter Reliability Program had oil on-site, and those are the plants that burned oil to keep the grid going. A chart on page 17 of the ISO-NE Cold Weather Winter Operations report[20] shows that more than 90% of the oil that was burned in the cold weather was burned at stations that participated in the Winter Program.

Oil saved the grid, and ISO-NE's foresight provided the oil.

Other situations on the cold weather grid

THE MOST IMPORTANT EVENT on the grid during the cold snap was the oil burn. Oil is what kept the grid going.

- Renewables kept up their usual level of input to the grid but could not make up for the lack of natural gas. As you can see in the chart on "Fuel Mix on the Grid," renewables were 9% of grid power before the cold weather, 7% during the cold weather, and 6% after the cold weather.
- LNG deliveries included delivering some gas from Russia to Boston. This delivery was after the cold snap. However, the Russian source led to questions and concerns because the United States has sanctions on some Russian fuel imports. However, the LNG importing company said that it was in compliance with US law and that they needed the gas to make up for the gas sold during the cold snap.[21]
- ISO-NE made many conference calls to other groups (pipeline operators and so on), which is part of their protocol for when the grid is stressed.

For further information on what was happening on the grid during the cold snap, I recommend the ISO-NE report on Cold Weather Operations.

For a shorter, more readable report, go to Rod Adams's blog, *Atomic Insights*. His post, "Performance of the New England power grid during extreme cold Dec 25–Jan 8," covers problems with the fuel supply and the disappointing performance of solar.[22] It also has an excellent comment stream.

Making a mountain of a small hill

Keeping the lights on in winter in the Northeast can be a major practical problem. In 2017–2018, it included LNG deliveries from Russia, emergency phone calls, and power plants nearly running out of oil.

This is clearly a mountain of a problem.

However, the constraints of the RTO system made that mountain. The initial problem was not quite a molehill; winter weather in the Northeast can challenge power systems. However, RTO governance turned a small hill into a mountain. Keeping the lights on in winter does not have to be quite so dramatic.

Electricity use is not constant. That is a simple fact. Electricity use rises and falls throughout the day and throughout the year.

A hot day with air conditioners means a high use of electricity. A frigidly cold night in New England also means high use of electricity. Many older homes in New England are not well insulated, and people supplement their main heat source (usually combustion of oil, gas, or wood) with small electric space heaters. Electric demand rises.

New England has comparatively few gas pipelines, and these pipelines supply gas to homes and to gas-fired power plants. On a cold night in New England, this leads to three problems.

- The homes have priority on the gas, and sometimes the gas-fired power plants cannot get gas in cold weather, when the homes are using a lot of gas.
- Power plants cannot store much gas on-site. The gas comes by pipeline. Any interruption in a gas pipeline means the gas-fired plants cannot produce gas-fired electricity.
- More and more power plants are dependent on natural gas, especially in New England, where 50% of our overall electricity supply is supplied by natural gas. In cold weather, lack of natural gas for power plants becomes a big problem.

The simple technical solution

Lack of natural gas for power plants can be a big problem in an RTO area.

In reality, the problem has a fairly simple technical solution. Many natural-gas power plants can burn oil or diesel or compressed natural gas as well as pipeline natural gas. If someone was in charge, in the RTO areas, the simple solution would be to order a certain number of gas plants to retrofit themselves to burn an alternate fuel and then to order such plants to stockpile that fuel. Their payments for the fuel stockpiles would go into their regulated rate base, and everyone would be happy.

Adding to the rate base in order to increase reliability was (and is) a common strategy in areas of the country where the utilities are vertically integrated. If power plants were encouraged to keep oil on-site in order to add to their rate base, they would do so.

In that case, the Balancing Authority would know that many plants would be available for operation, even in terrible weather. The plants would have a bigger rate base. Most of the problems would be resolved by the simple expedient of fuel storage on-site. Thank you, grownup in charge.

Oops, there isn't a grownup in charge in the RTO areas. There is nobody who can promise a power plant a better rate of return for being more reliable. And yet, ISO-NE did manage to set up a winter-reliability program that saved the grid by having oil stored on-site at some plants. How did ISO-NE do that?

It was uphill work for ISO-NE to set up the Winter Reliability Program, and FERC required them to dismantle it after three years.

Yes. It's a bit of a story, which illustrates some of the problems of the RTO areas.

CHAPTER 10

HARDER THAN IT SHOULD BE: PLANNING FOR WINTER

Winter reliability

TO UNDERSTAND THE Winter Reliability Program, we have to start with the fact that RTO areas try to do everything in a way that resembles a market. In chapter 6, "Vertically Integrated or RTO," we introduced the energy auction. In this auction, run by the RTO, plants bid in to sell their power (kWh). They bid in five-minute intervals. But the kWh auction is not the only auction run by an RTO. The RTOs set up many auctions, including new auctions to correct the results of older auctions. If there is a problem on the grid, an RTO will usually try to set up an auction to solve it.

For the winter-reliability projects, ISO-NE set up a new kind of auction. Dual-fuel power plants (power plants that could burn both oil and gas) bid in a special auction for the reliability program. They bid in their proposed costs for storing fuel on-site, planning to use the fuel when they could not obtain natural gas from the pipelines.[23] After the auction, ISO-NE picked the low bidders.

To most of us, the winter-reliability auction looks like a clumsy but effective way to get fuel storage without actually ordering a particular plant to store fuel. Such orders would be expected in a vertically integrated utility but are forbidden in an RTO area.

Name that fuel

SO, WE HAD OUR LOCAL winter auctions, with plants bidding to keep fuel on-site, except that the Federal Energy Regulatory Commission (FERC) objected. Everyone knew they would object because the ISO-NE had made the huge mistake of "name that fuel." FERC insists that ISOs must be fuel-neutral, so anything that smacks of "buy this fuel" is not allowed.

Really, though, ISO-NE was doing its best to be sure that it didn't completely "name that fuel" for the winter-reliability auctions.

Whether ISO-NE was or was not "naming that fuel" in the Winter Reliability Program, it had to put together a separate scheme to replace the Program. This scheme, required by FERC, would encourage enough power plants to keep fuel on-site, without actually saying, "Store fuel." Or even without writing anything that explicitly rewards plants that already store fuel by their nature (coal, nuclear, hydro, and oil plants might fit this description).

Department of Energy Secretary Rick Perry fell into trouble in the early days of the Trump administration. He was too direct in his suggestions for increasing the reliability of the grid. Perry asked FERC to ensure that plants that store fuel would be explicitly rewarded for storing fuel. There were other problems with Perry's plan. However, as far as I can tell, one part of the plan upset FERC and made two FERC commissioners (Robert Powelson and Cheryl LaFleur) state that Perry's plan would cause FERC to "destroy the markets."[24]

What was the part of Perry's plan that was so destructive? I think I know. Perry had broken the rules. Perry had "Named That Fuel."

Even before Perry was appointed Secretary of Energy, FERC had rejected the ISO-NE Winter Reliability Program as a permanent solution. To satisfy FERC, ISO-NE had to come up with a plan to encourage fuel storage without actually mentioning fuel storage.

Well, RTOs are not heavily staffed with lawyers for nothing! ISO-NE came up with a complex, clumsy scheme for Pay for Performance, in which plants that don't go online when called to do so have to pay part of their capacity payments to plants that come online. The scheme worked out so that the penalty payments would encourage plants to (let me whisper the words here) "keep fuel on-site" to avoid the penalties. We will talk more about the details of this plan later in the book.

Keeping fuel on-site is one of the ways that ISO-NE could provide winter reliability to the grid. But there is always conservation, and ISO-NE can encourage it. In terms of the winter-reliability planning, ISO-NE allowed for payments for conservation. Conservation took the form of demand response. A demand-response customer promises to go offline when the grid is stressed. Such a customer was allowed to bid into the winter-reliability auctions, just as oil-fired plants were allowed to bid. A demand-response customer would be paid for not using electricity, just as a power plant would be paid for keeping oil or LNG on-site.

Unfortunately for ISO-NE and the grid, there usually isn't a big response to demand response. In 2013, ISO-NE bought 1.9 million Megawatt-hours (MWh) worth of bids for the Winter Reliability Program. Only 4,000 MWh of demand-response customers bid into the auction. The rest of the bids were for oil on-site.[25]

This ratio did not change very much in subsequent auctions. In 2017-2018, ISO-NE paid approximately $24 million for oil as part of the Winter Reliability Program but paid only $34,000 for demand-response bidders.[26] ISO-NE just doesn't get that many demand-response bidders.

In short, very few groups want to go without power in winter. In fact, demand-response bidding functioned as greenwashing for the oil-storage program. Greenwashing is a form of marketing spin in which green marketing is deceptively used to persuade the public that an organization's products, aims, and policies are environmentally friendly. I do not mean to imply that ISO-NE planned the demand-response solicitation as a form of greenwashing. I am just pointing out that you can offer to pay customers to give up electricity on very cold days. However, very few will take your offer.

Pay for Performance

The winter of 2017–2018 was the end of the Winter Reliability Program, and future winters will be covered by Pay for Performance. The idea of Pay for Performance is that plants will be penalized if they don't go online when the grid is stressed. Hopefully, the fear of a penalty will encourage plants to keep oil on-site, without the RTO actually mentioning "oil."

It's almost impossible to understand the complex formula of Pay for Performance. For now, I will just say that ISO-NE had to account for several things in setting up Pay for Performance.

First, it had to define "when the grid is stressed." Plants go offline for maintenance, and so ISO-NE can't just slap a penalty on any plant that doesn't go online when called. Instead, the penalty depends on the state of the grid. The plant will be punished (fined) if it doesn't go online when the grid is stressed. If the grid is not stressed, and a plant cannot go online, ISO-NE will just call on some other plant. So, ISO-NE had to define "the grid is stressed" in terms of the Pay for Performance rule.

Next, ISO-NE had to define "go online when called." One of the issues has been that some plants (steam plants) ramp up slowly, and other plants (internal combustion) ramp up rapidly.

So, finally, ISO-NE was ready with its complex formula for Pay for Performance. ISO-NE has attempted to set up rewards and penalties that will work as well (they hope) as simply paying plants to keep fuel on-site. The complex set of payments and penalties is meant to force plants to go online in bad weather. These payments and penalties, carefully defined by the RTO, are supposedly an important aspect of a Free Market for electricity.

As a matter of fact, the entire RTO system is a complex set of payments and penalties that are supposedly in the service of a Free Market.

The RTO areas stretch the definition of "Free Market" until it pretty well breaks.

Having seen some of the complexities of keeping the lights on in winter in the Northeast, we need to look at some broader problems. How do RTO areas organize payments for electricity?

We will start with the breakups of various monopolies. Then we will look at the system as set up by the RTOs.

Warning. It is not a pretty sight.

POLICY, MARKETS, AND FUEL SECURITY

CHAPTER 11

BREAKING UP MONOPOLIES

Regulated monopolies

WE HAVE SEEN THAT RTOs make a mountain of a small hill: they are unable to take simple steps to ensure winter reliability. Later, we will see that, in the RTO areas, insider deliberations are quiet and closed. You need to be kind of a plutocrat to have your voice heard by organizations that run the RTO grid. How did this happen? What problems were RTOs going to solve?

Before RTOs, electricity utilities were "regulated monopolies." You were in the territory of one vertically integrated utility, and that was that. You were a ratepayer, with no choices.

As noted in chapter eight, the pre-RTO system of vertically integrated utilities had some major problems. The system had perverse incentives: it encouraged a utility to spend as much money as possible in order to get the highest profits it could obtain. The only barriers to spending were the rulings of the state's Public Service Board, and such boards were subject to regulatory capture. They could become way too close with the industry they were regulating. They could become rubber stamps for whatever the utilities wanted to do.

Meanwhile, in the late '70s and early '80s, the tenor of the times was very much in favor of breaking up monopolies and deregulating industries. Such actions had already been a success (more services, lower prices) with the breakup of the airlines and the telephone industries. These industries had also been regulated monopolies.

The breakup of the utilities might follow these leads. Indeed, the two monopoly breakups described above seemed to show a template for utility deregulation.

Airlines

Let's take the airlines for a start. In the '40s, the Civil Aeronautics Board (the main federal regulator) saw no particular reason that new airlines were desirable. The big four domestic airlines (United, American, Eastern, and TWA) took up all the spaces at the airports. Then there was Pan Am, the international airline. The situation looked well set. Airlines didn't compete on price, but they were allowed to compete on amenities. I remember my mother taking a rare plane trip. Before the trip, the airline called to ask how she wanted her steak prepared. Such questions were one of the major ways that airlines competed, since they all were required to charge the same price on the same route.

By the '70s, with hundreds of millions of people flying domestic routes, the system was breaking down. In consequence, airline routes and prices were deregulated. Almost immediately, new airlines started up, and ticket prices fell. Even more people traveled.

What lesson did the airline industry illustrate for the utility industry? There were two lessons:

- Deregulation leads to lower prices, better service, and more customers
- Deregulation does not lead to more accidents. The safety-regulation part of the regulatory scheme stayed in place and perhaps was even enhanced.

Since utility operations have the potential for all sorts of accidents, from live wires falling on a road to explosions at fossil and nuclear power plants (and let's not forget dam failure), the improving safety record of the airline industry was a good argument for utility deregulation.

Airline deregulation showed that deregulation did not have to affect safety.

Telephones

Not many people today remember the strict monopoly powers of Ma Bell, a monopoly that mostly ended in 1982. Long-distance calls were madly expensive, calling from a hotel was a mini-nightmare, where the hotel switchboard patched you through to an AT&T long-distance operator.

Not content with owning the phone lines, the phone company also owned your phone and charged you more if you had an extension phone in your house. Some of my law-abiding friends realized that the way the phone company knew you had an extra phone in your house was through feedback from the ringing mechanism. Bootleg phones abounded: the upstairs phone didn't ring (its ringer was gone), but at least it was there. When you heard the downstairs phone ring, you could at least get to the upstairs phone without a mad dash down the stairs.

The phone company rules turned some of my law-abiding friends into criminals, altering their phones to avoid extra charges. Not me, though. I wasn't a criminal. And not because I am a goody-two-shoes. We just never had a house big enough for an upstairs until long after the phones were deregulated. George and I married while we were still undergraduates in 1965. The landmark phone-deregulation case was in 1982.

Deregulating the phone system led to lower costs, more choices for long-distance carriers, cell phones all over the place, and all sorts of good things for the consumer.

In some ways, deregulating phones seemed an impossible task. These were the days when everyone had a landline, and there were no cell phones except for specialized uses. For example, in the early '80s, I rode in a utility truck to a geothermal field and was very impressed with the phone in the truck. Such phones helped the utility keep track of its people, and they could also connect to the regular phone system. But unless you were in a utility truck or a police cruiser, you were not connected with the phone system while moving.

So back to deregulation. The immediately perceived problem with breaking up the phone monopoly was "What are they going to do? Have four utility poles near the house, with four possible wires into the house for the phone?" In other words, how could competition work when there was going to be only one line into the house?

What happened was a bit of a kludge, but it worked. People had only one line into the house, but they had many choices for their long-distance carrier. The carriers also offered many types of plans, and you could choose according to your phone-usage patterns. Overall, the cost of long-distance phone calls dropped rapidly. In other words, even before cell phones became common, breaking up the telecommunications monopoly improved the choices for the consumer.

Phone deregulation showed that there could be significant consumer benefits to deregulation, even with one line into the house.

Electric utilities

And yet, and yet ... something went wrong with utility deregulation. Careful analysis may show some savings for consumers in the deregulated utility areas, but a first-pass look at the situation shows no savings (see figure 6). You have to look closely and analyze hard to see any savings in the deregulated areas, a situation quite unlike airlines and phones.

With airlines and phones, any person who was an adult under regulation and deregulation would say, without deep analysis: *Yeah, things are better and cheaper now. Better phone service, more flights.* This is not true for utility deregulation. RTO areas have higher prices.

Figure 6 is from a paper by Severin Borenstein and James Bushnell of the Energy Institute at Haas (University of California at Berkeley).[27]

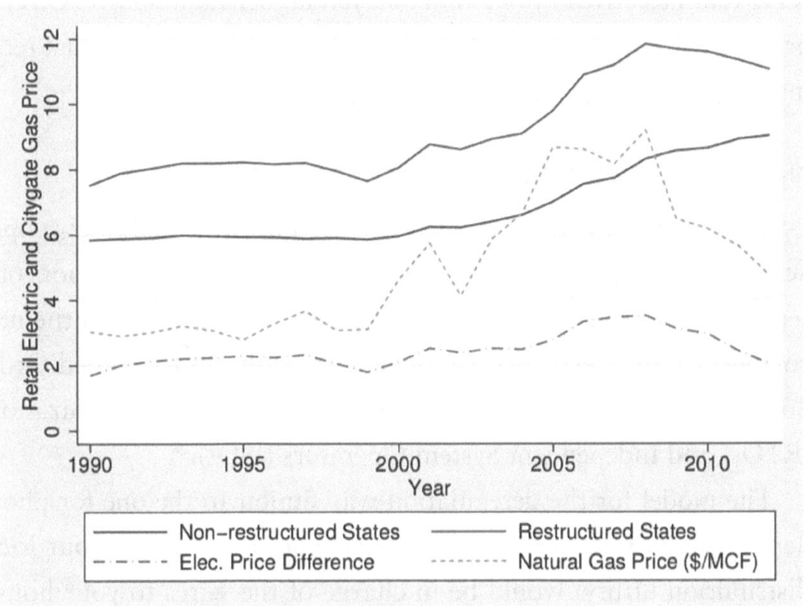

Figure 6: Comparison of regulated and RTO states ("restructured states") (Borenstein and Bushnell)

They distinguish between the "restructured states" (RTO areas) and the non-restructured states (states with vertically integrated utilities). It is easy to see the restructured states: they are the line near the top of the chart (highest cost per kWh retail rates).

In their paper, Borenstein and Bushnell do not conclude that electrical prices in the RTO states are more expensive due to restructuring. While the "restructured states retail cost" line on the chart starts high and stays high, the authors point out that restructured states had electrical retail costs that followed the gas prices more closely than the retail costs in the non-restructured areas. (The line for gas prices is the dashed line with the small dashes.)

While the "restructured states" with RTOs remained more expensive than the non-restructured states, the gap is narrowing. However, whatever the authors of the paper conclude, I would point out that RTO retail electricity prices have stayed higher than non-RTO retail prices for twenty-five years. I would not personally conclude that retail markets have saved money for the consumer.

Utilities and FERC orders

In 1997, FERC issued Orders 888 and 889, designed to give open access to transmission lines, and to "encourage the creation of a separate Price Exchange to reveal market-clearing prices in the new competitive market." Two years later, in 1999, FERC issued Order 2000 to foster participation in Regional Transmission Organizations (RTOs) and Independent System Operators (ISOs).[28]

The model for the deregulation was similar to the one for phone deregulation. One service (your local phone company, your local distribution utility) would be in charge of the wires to your house. Meanwhile, there would be competition on the bigger scale, such as long-distance carriers and power plants.

The "one service that is in charge of local wires" and "choices of other services, like long distance" became part of the model for utility deregulation.

Utility deregulation

THE LOCAL UTILITY WOULD still manage the wire into your house and take care of billing and downed lines during storms. However, you would have your choice of energy suppliers (the local utility is in charge of distribution). You could base your choice on your own criteria. Green power? Inexpensive power? Cheap power up to this much usage with penalties for more usage? Companies could offer it, and you could choose it.

Oops. I forgot something. This didn't happen, or, at least, it didn't happen very often. The consumer was left out of the equation. FERC issued orders that allowed local areas (not well-defined) to set up a market system (also not well-defined). Some set up market systems; some didn't. That is how we get the multi-shaded map of RTO and non-RTO areas, shown in figure 3.

In general, areas of the country with high electricity rates or few publicly owned utilities set up some kind of market system. Perhaps they hoped to save money; perhaps the investor-owned utilities that dominated those areas hoped to make even more money than they were making. I don't know. I do know one thing. It is not clear whether, twenty years after the FERC order, areas that had set up a market system had saved any money whatsoever.

What went wrong with utility deregulation?

WHY DID AIRLINE AND PHONE deregulation work for the consumer, but not utility deregulation? To some extent, this entire book is about what went wrong with utility deregulation, so it is tough to put together a short summary. But a few things stick out as major problems:

- **No consumer choice:** In airline and phone deregulation, consumers were given choices. Sometimes the choices were broad (new airlines, new fares, new routes) and sometimes narrow (long-distance carriers, minutes per month). But the consumers always exercised more choice after deregulation than before deregulation.

  In contrast, very few utility ratepayers were given any choice after deregulation. Ratepayers are still ratepayers. Prices didn't fall after supposed deregulation. Without consumer choice, prices can't be expected to fall. Much of the problem with utility "deregulation" is as simple as that.
- **No transparency:** The stakeholders (insiders) meet in secret or semi-secret. They all have very good economic reasons to use their full power to influence which power plants are on the grid and how those plants get paid. Mere ratepayers will never learn what "stakeholders" share with each other.
- **No accountability:** Nobody is accountable for the grid. A power-plant owner can run his power plant well or badly: it's not his plant's fault if the grid can't get enough power. Just-in-time natural-gas plants can take over a grid, and just-in-time deliveries can fail in cold weather. It's not the RTO's problem to keep a mix of plants on the grid, for resilience. In RTO areas, generation utilities own power plants, and distribution utilities own power lines and substations. Distribution utilities still have regulated rates of returns and have to keep their lines and substations in good shape. Generators do not have that level of responsibility. The RTOs that manage the whole system do not have a high level of responsibility. Wherever you are in

an RTO area, unless you are a distribution utility maintaining local lines, the buck never stops with you.

Before moving on to an in-depth review of the problems in the RTO areas, I want to quote from an essay by two utility lawyers, which was published in *Utility Dive* in November 2018. Ray Gifford and Matthew Larson wrote "For RTOs & ISOs: 'Don't call it a market' (props to LL Cool J)."[29] The essay describes how the term "market" for RTO areas is "brilliant PR" and why the elaborate "amalgamation of political fixes and regulatory band-aids" in RTO areas is not a market.

I couldn't have said it better myself.

CHAPTER 12

GAMING THE SYSTEM

Generation utilities and distribution utilities

In an RTO area, distribution utilities buy electricity from generators. Generation utilities generate electricity.

Generation utilities are not really utilities. They are power-plant owners who sell into the wholesale markets, and they run their plants as long as they see the plants as being profitable. They are not guaranteed a rate of return, and they are often called "merchant generators."

The distribution utilities buy power on the wholesale markets, and they sell the power to their customers. We will describe the wholesale markets in a future chapter, but, for now, let's just look at the reality of one of the first wholesale markets. At the turn of the century, California had rolling blackouts. Power was turned off in one area and then turned on while the outage "rolled" on to the next area. Yes, elevators got stuck, traffic lights failed, and general confusion ensued.

How did this happen?

Nobody is responsible

In California, as elsewhere, the distribution utilities were responsible for keeping their lines and substations running. The lines and substations ran fine. The RTO (CAISO) was responsible for dispatching the plants that were available for dispatch. If too many of them were offline for "maintenance" and there had to be "load shedding" (rolling blackouts), well, CAISO is not responsible for plant maintenance, just for dispatch.

Every individual plant makes money when it runs. If a plant needs maintenance or whatever at an awkward time, it's not making money at that time. That individual plant is not responsible for keeping the grid going. No individual plant owner is responsible for the grid.

Actually, the RTO isn't responsible, either. All sorts of laws were passed, supposedly to protect the consumer. The distribution utilities had to buy electricity on the wholesale market. However, the legislature put a cap on the prices that the distribution utilities could charge the consumer. The cap held fast, no matter how much the distribution utilities had to pay for the electricity on the market.

Gaming the system is the name of the game

The system was set to be gamed. Opportunities arose. Enron, Shell, and NRG owned several plants that fed into the California grid. They closed plants for "maintenance" in order to create scarcity and drive up the wholesale electricity price. Their other plants received the new high prices. Lack of generation led to rolling blackouts in California. These "maintenance" closings were a major reason for the blackouts in California in 2000–2002.

In fairness to Enron and the other merchant generators, their manipulations were not the only reason for the blackouts. California regulators had told the distribution utilities that they had to pay market costs but would be limited in what they could charge customers.

Their costs could rise, but their revenues could not rise. When some of the distribution utilities began losing money and being late with their payments to generators, some generators ran out of money or patience for extending credit. They simply didn't buy fuel to run their plants or produce power, since they were not sure they would be paid in a timely fashion.

Meanwhile, the state of California knew the markets were being manipulated. They complained to the RTO but got no help from that source. The state then complained to FERC, the federal regulator. FERC also offered no help whatsoever.

In some ways, this was a bigger version of the problem on the New England grid in the frigid weather. In New England, there was nobody who could order a plant to keep oil on-site. In California, there was nobody who could order a plant to go online if it claimed it needed maintenance. An RTO can't "name that fuel" and can't "name that plant." In New England, grid fragility arose because nobody could order a plant to be ready to run. In California, the fragility was partially due to the system being gamed. In both cases, it was the RTO rules, not natural disasters, that led to the problems.

In 2015, a federal appeals court affirmed that the California market had been manipulated in 2002 and that nobody (including FERC) had stopped it. *The Sacramento Bee* carried an opinion piece about this ruling, including the unanswerable question: What took everybody such a long time to do something about this?[30]

Market manipulation was successful in California because there was nobody responsible for making the system work. In an RTO area, the buck *never* stops *anywhere*. Not even today.

CHAPTER 13

THE DEATH OF THE MARKET HOPE

Who feeds Paris?

In theory, if electricity is really a market, then nobody needs to take responsibility for how it operates.

There's a famous metaphor about the difference between markets and planning: "Who feeds Paris?" Who does the planning that ensures that Paris will have enough food every day? Is there a central planning agency at the Hôtel de Ville? Of course, there isn't. The market feeds Paris. There is no central planning authority.

Every boulangerie, patisserie, charcuterie, brasserie, and bistro orders the food that it thinks it can sell. Each chooses flour, butter, and meat based on quality, cost, and what its customers want. There is no central planning, but, every day, Parisians eat the food they prefer, from the vendors they prefer, at reasonable prices for the quality they prefer. Because, in "Who feeds Paris?" the market does better than any central planner could possibly do.

That was the hope of the RTO areas. But the hope was not fulfilled.

The RTO areas are not a market.

Travis Kavulla (at the time, he was a Commissioner in the Montana Public Service Commission) wrote an article titled "There is no free market for electricity: Can there ever be?" This erudite article was published in the summer of 2017 volume of *American Affairs*.[31]

Kavulla writes: "Paradoxically, competitive markets appear to attract more regulation.... The tariffs (rules) that govern the competitive ISOs are approved by the Federal Energy Regulatory Commission (FERC) ... (these) tariffs governing 'market' relations are enormously complex, spanning several times the length of the tariffs that govern the vertically integrated monopolies ..."

In other words:

The RTO areas are more heavily regulated than the non-RTO areas. They are not markets as we know markets. They are complex systems, with new regulations constantly tweaking and trying to improve existing regulations. They are a bureaucratic thicket, not a market.

It's Orwellian. RTOs are "deregulated" only if "deregulated" actually means "lots more regulation."

"War is peace." "Deregulation" is "lots more regulation." Orwell would be amused.

Regulations raise the costs

ARE THE ELECTRICITY customers amused? No. They usually don't have a choice of vendors. Utility customers are still merely ratepayers. Meanwhile, RTOs and their regulations cost money for the consumer. (More about customer choice in chapter 39 on Manipulating the Customer.)

Four studies show that RTOs cost money for the consumer, but only three studies were willing to actually draw this conclusion.

Borenstein and Bushnell 2015

Severin Borenstein and James Bushnell of the Energy Institute at Haas (the business school of the University of California Berkeley) tried to tease out the effects of RTO areas on consumer prices.[32] One of the key graphs from their paper is reproduced as figure 6 in chapter 11. RTO areas had somewhat higher prices throughout the twenty years that were reviewed, but the authors concluded that it was unclear whether the prices were due to the RTO structure or even statistically significant.

The authors admit higher prices for the consumers in RTO areas but are unwilling to conclude that the prices are due to the RTO structure.

Marcus 2011

In 2011, William B. Marcus, who was then chief economist for JBS Energy in West Sacramento, did a careful comparison of RTO areas and traditional areas, using two methodologies to compare different areas of the country and take account of different fuel mixes.[33]

Marcus concluded that being in an RTO area increased consumer costs by 1.25 to 1.5 cents per kWh.

Gattie 2018

David Gattie of the University of Georgia compared residential rates in regulated and RTO areas.[34]

In his graph of residential rates in regulated areas from 1980 to 2016, skipping Vermont, the residential prices range between 4 and 11 cents per kWh, clustering around 8 cents in 2016. In the "deregulated"

RTO areas, retail prices are between 9 and 21 cents per kWh, with most prices between 11 and 19 cents per kWh.

In the regulated-area graph, Vermont's high prices stick out at more than 15 cents per kWh in 2016. However, Vermont is partially deregulated and participates in the RTO auctions. Vermont costs would fit better with the costs in the RTO-area state graph.

However Vermont ends up being classified, the consumer prices in the RTO areas are clearly significantly higher. As Gattie concludes: "If there's an advantage to deregulated, non-vertically integrated utility markets ... rates aren't one of [the advantages]."

Dyer 2019

R.A. "Jake" Dyer is a policy analyst for TCAP, the Texas Coalition for Affordable Power. His organization sells electricity to municipalities in Texas.[35]

When lawmakers in Texas deregulated electricity in 2002, they allowed municipalities to remain regulated markets with a single provider. Therefore, Texas has RTO areas and also non-RTO areas.

In the TCAP report "Electricity Prices in Texas, A Snapshot Report, 2019 Edition,"[36] Dyer compares the price of electricity between the RTO areas and the non-RTO areas within Texas. Figure 7 shows the comparison during the deregulated years. He shows that the prices in the deregulated areas of Texas soared after 2002 but have recently come down to approximately the same level as the regulated areas. He uses data from the United States Energy Information Administration.

Dyer estimates a quantity that he calls "lost savings," the extra money that Texans in RTO areas paid for power. As he writes: "'Lost savings,' as defined in this report, is the imputed savings that would have accrued to Texans living in areas of Texas with deregulation if they paid the same average prices as Texans living in areas exempt from deregulation." The extra costs of the RTO areas are staggering.

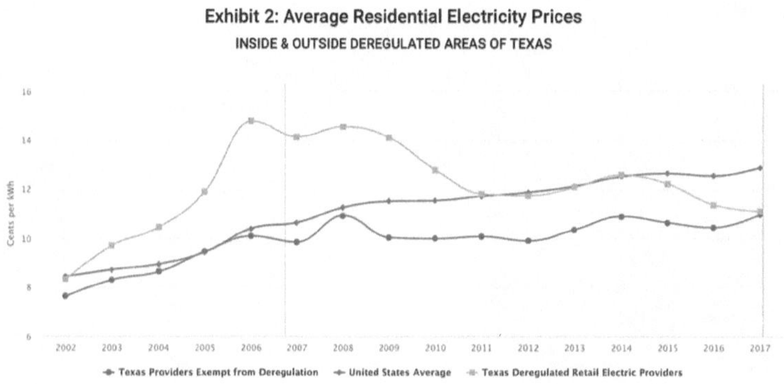

Figure 7: Electricity prices in Texas, deregulated and regulated areas (Dyer, TCAP)

Texans in RTO areas paid a total of $27 billion more, for the years 2002 to 2017, than they would have paid if they were in non-RTO areas. This is an extra cost of approximately $5,500 per household.

To summarize, all four reports show that deregulated areas have higher consumer prices than vertically integrated areas. However, because electricity pricing is a very complex field, only three out of four authors were willing to take the next step and ascribe the higher prices to the RTO system itself.

Nobody claims that the RTO areas save money for the customer. That much is clear.

CHAPTER 14

HOW POWER PLANTS GET PAID

Now that we have seen the effects on the consumer, let's look at the policies that affect the power plants and at the strategies that the plants use to respond to those policies.

We have to start with the fact that not all power plants run all the time. In the old days of vertical integration, this wasn't much of a problem, because the utility was happy to build and maintain more power plants than were needed. All those plants went into the "rate base" and increased the utility's rate of return. However, as soon as the RTO auction process was underway, the power plants were owned by a variety of entities, and the owners had to make their decisions one power plant at a time.

Starting with the auction

At the first level, in the RTO auction process, the power plants are paid per kWh produced, in a series of real-time auctions and day-ahead auctions. The plants could also make Power Purchase Agreements (PPAs) with a distribution utility, and many did.

PPA terms do not have to be made public, and probably more than half of electricity is sold by those agreements. Therefore, some people question whether the auction prices have very much effect on the wholesale price of electricity.

I think the auction prices determine most prices on the grid. Of course, I can't prove this, since the terms of the PPAs are not publicly available. But I look at the incentives. Both sides entering into a PPA want to be sure that the prices in the PPA are pretty darn close to the prices in the grid auctions.

Nobody wants to feel cheated: "I coulda' bought it cheaper from the grid!" Or, alternately, "I coulda' made more money selling directly on the grid!"

Since all the PPA players had incentives to keep the prices close to grid prices, and PPA agreements are private and seldom made public, I will continue to write about the grid auctions. I also need to mention that some generators "self-schedule." With the permission of the grid operator, such generators commit to being on the grid for certain hours. The self-scheduling generators receive the clearing price for their kWh. Once again, the clearing price is most important.

The auctions set the clearing prices, and the clearing prices influence the PPA prices and set the prices for self-scheduled units. The clearing price basically sets the prices on the grid.

The power-plant owner

Let's examine the incentives for a power-plant owner in an RTO area. The owner has to decide: Will this power plant run often enough to generate enough revenue for its upkeep? Will it be a hard winter (my power plant will run a great deal, and I will make lots of money) or a mild winter (maybe not so good for me)? How much will my fuel cost? What will be the price on the grid?

In other words, with complete uncertainty as to revenues, each power-plant owner had to make a decision about paying for maintenance of the power plant. If an owner has more than one plant, the owner must decide separately on the fate of each plant. Will that plant make money? Should I shut it down?

All these decisions looked dangerous for reliability. Companies could just shut down their power plants. Too many power-plant owners could decide that their power plant wasn't going to be paid enough, and then there wouldn't be enough power plants on the grid. Or power-plant owners could game the system by shutting down power plants (as they did in the early days in California).

The capacity auction

SOME OF THE RTOs DECIDED that they needed another type of market in order to ensure that there were enough power plants available when needed. The RTOs looked at it this way: the plant is not only selling kWh, it is also selling something else the grid needs. It is selling power-plant capacity. It is selling its availability to make power when the grid needs it.

In many RTO areas, plants sell "available capacity" whether or not they are making power at the moment. Power plants get capacity payments, and they can use these payments to maintain their plants.

ERCOT, the Texas RTO, and CAISO, the California RTO, did not implement such capacity payments.

As I write this, ERCOT is facing very low reserve margins this summer.[37] (Reserve margins are the number of MW available above the number of MW expected to be required, in other words, MW "in reserve.") Low reserves will lead to higher prices for electricity and may lead to not having enough capacity to meet demand.

In contrast, California expects to have sufficient margins, partially due to abundant hydro energy from a generous snowpack.[38]

Capacity payments are supposed to encourage power plants to be available and therefore stabilize the reserve margin. Looking at California and Texas, it does not seem that capacity payments are make-or-break for reserve margins. But once an RTO had capacity payments, then the door was open for many more types of payments and auctions.

Indeed, it is time to learn more about day-to-day management in the RTOS.

Economic dispatch and auctions

THE BASIS FOR RTO GRID management is economic dispatch and auctions. Let's start with the energy auction, the five-minute auctions where plants bid to provide kWh.

Economic dispatch is pretty much just as it says: the least-expensive plants are dispatched first. If the grid needs 1000 MW of power, and a low-priced plant can meet that demand, the low-priced plant will be dispatched to meet it. If the grid needs more power, and the only plants remaining are higher-priced, then they will be dispatched next. In general, this assures that low-priced plants will run more hours than high-priced plants, and the price of electricity will be kept relatively low.

Utilities have always chosen economic dispatch in order to keep their rates down. Under vertical integration, it didn't matter very much which plants a utility dispatched, since it usually owned all the plants anyway. Still, even though the utility was allowed a rate of return on their investment, the PUC tried to ensure the utility would attempt to keep the customer rates as low as reasonable, considering the plants in their fleet. As we can see by reviewing chapter 13, "The Death of the Market Hope," the PUCs have been good at keeping prices low in the non-RTO areas.

Back to dispatch. Least-expensive-first wasn't (and isn't) the only constraint. There are physical constraints also. All other things being equal, the dispatcher will choose to dispatch plants that are near the load so there won't be excessive line loss. The dispatcher must also keep track of how much electricity a line can carry, and not overload any of the lines.

For simplification, however, we will assume that RTO dispatch is driven completely by economics. In RTO areas, the economics of dispatch is set by auctions.

Three shocking things about the auctions

Shocking thing the first: clearing price

When I am teaching my grid class, I can always get a gasp out of the crowd by explaining the basic auction.

"So," I explain, "I'm the RTO, and I need 500 MW right now, or rather, I need it for the next five minutes. The auction runs every five minutes."

> Power plant A steps up: I've got 200 MW for you, at 15 cents per kWh.
>
> Plant B steps up: I've got 100 MW for you, at 20 cents per kWh.
>
> Plant C steps up: I've got 300 MW for you, at 30 cents per kWh.

I'm the RTO, and I answer: "Okay, plant A and plant B, I'll take all your output. Now I have 300 MW. Plant C, I'll buy 200 MW of your output, but I don't need all 300 MW. I've got my 500 MW now.

"All you plants, you get 30 cents per kWh for your output. Plant C has set the clearing price for this round."

At this point, the people in the class begin to shake their heads:

"So plant A bid in at 15 cents and is getting 30 cents?"

The answer is: "Yes, indeed." That is how the auctions go.

This "clearing price" system is usually considered the incontrovertible method by which ISOs run. I want to point out that not everyone is happy with this scheme. "Clearing Price" is not written in the heavens or in the laws of physics. It is just how the ISOs do things. It could be changed.

Shocking thing the second: RTO control

The RTO controls the bids. They know your fuel costs. They know your heat rate, which is how efficient your plant is at turning fuel into electricity. They can calculate the cost of your next kWh of electricity just as well as you can.

They keep track of your costs and make sure nobody is bidding too much.

ISOs will tell you that nobody wants to bid too much, because bidding too much means not being chosen. Bidding too much means your plant won't sell its power. But it doesn't matter, really, what you want to bid. The RTO knows the marginal cost of the next kWh your plant makes, and that is what you are going to bid. They are checking.

Sorry, but there is a complication, of course. You can add to the bid price under certain circumstances. You can add a "risk premium,'" but the amount you can add is limited by RTO rules.

(Yes, I hate writing about these things, because they are always rules on top of rules. For this iteration, just realize that, basically, the RTO controls the bids at the auction.)

Capital costs are not allowed in your calculation of how much you bid. However, outside sources of income are allowed and welcome.

If you are paying off a high-priced plant, you can't add any of that expense into your energy bid. But if you are receiving payments for Renewable Energy Certificates, you can take those payments into account and bid at a lower cost for your kWh.

RTOs claim to be fuel-neutral, therefore, all expenses and income streams for all types of plants are equal in the eyes of an RTO. But some are more equal than others.

Shocking thing the third: payment according to their costs

The payment money depends on other people's bids.

To some extent, this is always true in markets. If you want to grow zucchini for profit, you have to factor in both the cost of your land and how productive it is, as well as the current selling price of zucchini. If you paid a lot for the land but it is very productive, and if zucchini is scarce and/or popular, you can name your price to make a profit.

Setting a price for electricity in an RTO area is more complicated than setting a price for zucchini. You are going to have to compete separately on capital costs and production costs, and your bid is sliced, and it is diced. You have a kWh bid and a capacity bid and maybe an extra capacity payment bid (Pay for Performance, and maybe another payment for ancillary services). You don't get to decide on any of these bids. And the RTO will be looking over your shoulder every five minutes to be sure you do it right.

The role of the clearing price

In an RTO energy auction, power plants get paid the clearing price for those five minutes. They get paid the clearing price, no matter what they have bid. Being paid the clearing price is a decision made by the RTO: there is no law against paying plants the amount they bid, instead of paying them the clearing price.

When I think of it, there is not much "usual justification" for using the clearing price as a payment metric. There is an idea that the generators, if chosen to run, can expect to be paid their marginal price (if they are the highest bidder and set the clearing price) or more than their marginal price (if they are a lower bidder and receive the higher clearing price). This system doesn't actually help steam plants that much, because when renewables are forcing the price on the grid down to one cent, zero cents, or minus one cent per kWh (we'll pay you to take our power), steam plants often choose to keep running and take the price, far below their operating costs, for several hours a day.

The main explanation for Pay at the Clearing Price is that the clearing price is attached to the marginal cost (for the highest-priced plant), while Pay as Bid would mean that utilities would bid in at the price they hoped and expected to get, which would be more than their marginal costs. So, the bids would have some randomness that the RTO would not have a method to control. The plants would control their own bids.

Exactly this sort of thing happens in real markets. Everyone tries to outguess and outcompete everyone else. Naturally, those in charge of the electricity "markets" hate this. What would happen to the thick books of tariffs in the RTO areas, if everyone could just bid into the market as they chose?

The supposed fear is that "Pay as Bid" would lead to higher prices.

In fairness, that could be an issue for power plants, where there are just not that many bidders (unlike the hundreds of farmers that could supply zucchini or butter to Paris). Some papers on Pay as Bid conclude that the generators would bid the price they hoped to get, and the overall pricing of the market would be higher. Conceivably, Pay as Bid might lead to higher prices for consumers. The fear is that big generators would have near-monopoly power.

The New York Independent System Operator, the RTO for New York State, commissioned consultants to study Pay as Bid pricing.[39] A version of this paper was published in the March 2008 edition of *Public Utilities Fortnightly*, but it is behind a paywall at that journal. The authors conclude that "pay as bid" would not save money or increase reliability. The following discussion is influenced by this paper.

There are relatively few big generators on any grid, but they don't have monopoly power. Grids have a comparatively small number of large plants that run all the time. Most grids have a great many power plants that don't run all the time but come on the grid to follow load. In my opinion, these part-time plants could undercut a large generator that was charging too much and acting like a monopoly.

In real markets, after a while, everyone learns how to bid properly to make a living. Also, real markets tend to be a little cutthroat, which tends to benefit the consumer.

People who describe how Pay as Bid in electricity wouldn't work well often argue that "reducing the exercise of market power" is a good result of paying the clearing price, instead of paying at the bid prices. Market power is generally defined as the ability to raise prices without losing customers. That is, market power is akin to holding a monopoly. Market power is being a "price maker" rather than a "price taker."

However, there are already price makers and price takers on the RTO grids. RTOs sometimes make rules that certain types of plants can be only price takers, not price makers. In a recent ruling to save the Mystic Generating Station in New England, FERC agreed that Mystic would be paid for cost of service, including a return on investment. In other words, Mystic will be paid as if it were a regulated utility in a vertically integrated market. However, Mystic must bid into the auctions at $0 in order to avoid raising the price on the grid to the cost of its service price. Within the auction, Mystic will be a price taker. However, Mystic's bid at zero will probably lower the clearing price on the grid, which will not make the other power plants happy. The agreement with Mystic was well described in an article by Sarah Shemkus in *Energy News*.[40]

As we can see in the Mystic example, FERC and the RTOs can manipulate the "market power" issue, including how the clearing price is set. And they do so.

In fact, RTO areas pay by clearing price, and these areas have higher prices for the consumer than non-RTO areas. Still, I am sure that RTO area managers will dispute the idea that paying the clearing price raises the price.

Auctions work for the RTOs. I am not at all sure they work for the consumer.

CHAPTER 15

SELLING KWH IS A LOSING GAME

Who gets paid for what?

IN EARLIER CHAPTERS, I described kWh auctions and mentioned capacity auctions. Different types of power plants have different strategies, depending on which set of auctions they rely on. When we understand who gets paid for what in the auctions, we can predict the strategies of the various types of electricity providers in the RTO areas. As the rules in the RTO areas get thicker and thicker, the plants spend more time figuring out their strategies.

In the long run, RTO markets punish reliable plants and support unreliable plants.[41]

When Entergy announced that it was going to shut down the Pilgrim nuclear plant, it explained its choices by preparing a chart of revenue streams. I later used that chart in an article in *Nuclear Engineering International.*[42]

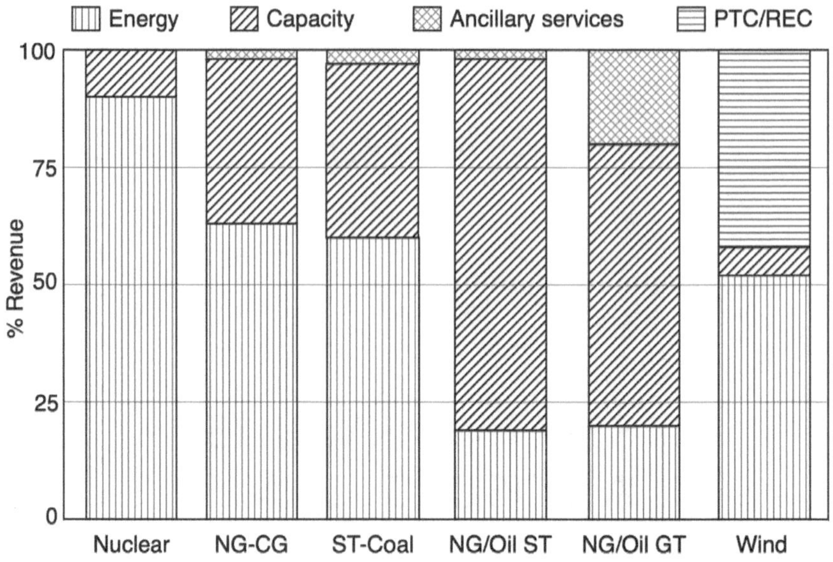

Figure 8: Revenue streams for different types of plants *(Nuclear Engineering International)*

This chart describes the revenue streams for several types of power plants on the New England grid. "Nuclear" is nuclear, of course. NG-CC is natural gas, combined cycle, which is another type of plant that is often used as baseload. ST-Coal is steam-turbine coal. NG/Oil ST is a plant that uses natural-gas or oil and uses a steam-turbine cycle. NG/Oil GT is NG/Oil-Gas Turbine, which is a plant that can use natural gas or oil and uses a gas turbine (internal combustion, no steam) cycle. Wind is wind, of course.

We can see that nuclear plants get around 90% of their revenue from selling kWh (energy), while gas plants get 40% to 80% of their revenue from other sources (capacity payments, ancillary-services payments). If gas plants had to subsist on the prices they received for energy, that is, if their main income was the money they receive for selling kWh on the grid, it would be much harder for them to undercut nuclear pricing.

Three strategies

LOOKING AT THE REVENUE streams, we can see how the various strategies will shape up:

- Nuclear plants and coal plants make a lot of power and live by energy prices. They need a reasonable high price "on the grid" (energy auctions) to survive. There are very few coal plants left in New England, so this is mostly about nuclear plants.
- Gas-burning plants get much of their revenue from capacity payments; they will fight anything that threatens those payments. Later, in the sections on MOPR (Minimum Offer Price Rule) and CASPR, (Competitive Auctions with Sponsored Policy Resources), we will see changes to capacity payments. We will also see the gas plants insisting that various changes to capacity payments are "breaking the market." Nobody would want to break the market, would they?
- Renewable plants live and die by renewable-energy-certificate payments and production tax credits (REC and PTC). They will fight to defend these. Renewables are also fighting to get higher capacity payments (see CASPR), putting them a little bit at odds with the gas-fired plants. However, CASPR also rewards gas-fired plants. So, "it's complicated," as they say of relationships on Facebook.

Capacity is the winning strategy

How are these strategies working out? Who is winning?

The gas-fired plants are winning. Capacity prices are up, and energy prices are down. This is good for the gas-fired plants.

Figure 9 shows New England energy (kWh) and capacity prices between 2008 and 2018. Over those ten years, the amount of money being paid for energy has stayed steady or drops, but the amount of money in capacity payments continues to rise.[43]

- In 2009, energy payments were $5.9 billion, and capacity payments were $1.8 billion.
- In 2018, energy payments are estimated at $6.0 billion, while capacity payments are estimated at $3.6 billion
- The depend-on-capacity-payments strategy is a winning strategy for a power plant.
- The depend-on-selling-kWh strategy is a losing strategy.
- Depending on RECs and PTC (not shown in figure 9) can also be a winning strategy.

These statements sum up a lot of the issues in the RTO areas.

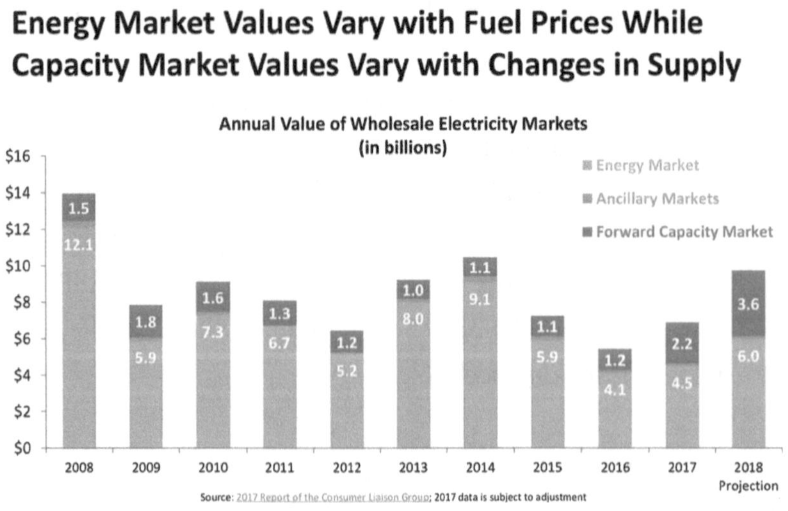

Figure 9: Annual value of New England wholesale electricity markets (ISO-NE)

As usual in RTO areas, it's more complicated than that (day-ahead auctions, MOPR, CASPR, and so on). We will explore these complexities and the subtle strategies that give rise to the complexities.

Still, it's worth thinking about the fact that, in most RTO areas, it's not a good strategy for a power plant to depend on selling kWh in order to make a living.

CHAPTER 16

GOVERNANCE AND DRAMA

Keeping the lights on

In chapter 10, "Harder Than It Should Be: Planning for Winter," we looked at how ISO-NE is hamstrung by the requirement to appear "fuel-neutral." That requirement led to a great deal of unnecessary drama during frigid weather in the Northeast. ISO-NE could not simply order plants to keep oil on-site, because that would have been "Name That Fuel." Consequently, plants were running low on fuel, as natural gas was used for home heating.

The way the RTO grids are governed leads to most of the crises and the drama on the grid. The plants, all of them, just do the best they can.

I described some of this material in the earlier section "An easy problem." That section discussed ISO-NE, the grid operator, as if it worked alone. In this section, I will continue to use the frigid weather in New England as the example, but there will be many new characters in the drama. I will look at the roles of various characters (FERC, ISO-NE, NEPOOL) and the final plans for reliability of the winter grid. I'll start by asking a simple question: How come most people don't know much about the grid itself?

No Sunshine Laws on the Grid

WHY DON'T PEOPLE know much about the grid?

I used to think that consumers didn't know much about the grid because they didn't care. Another possibility is that power plants are easy for people to explain: I love/hate nuclear. I love/hate solar. And so forth. Grid choices such as payments on the winter-reliability projects—that's hard to explain. It is hard to know the situation well enough to even have an opinion.

However, I have concluded that the reason people don't know much about the grid in the RTO areas is because people can't find out what is happening on the grid or who is controlling it.

- The RTO areas have extraordinarily complex rules. Only insiders can follow the complexity. In contrast, in vertically integrated utility areas, an ordinary person can go to a state PUC hearing and figure out what is going on.
- The RTO areas often actively discourage public participation. My own power pool (New England Power Pool) bans reporters from its important meetings. Other RTOs don't explicitly ban reporters, but their rules are so complex and their meetings are so lengthy that only industry journals (*RTO Insider*, *Utility Dive*) report on their activities. Only "insiders" go to the meetings.

People don't know about the grid because the RTO areas do not encourage transparency.

So how did we get this collection of insiders who run the grid in closed meetings? As usual, I will use my local grid as an example.

As described on its history page,[44] in 1971, the New England Power Pool (NEPOOL) was founded to coordinate various aspects of power supply in New England. NEPOOL is a consortium of energy insiders (aka "stakeholders"). Its most important subgroup is the "Participants Committee."

In 1977, the Federal Energy Regulatory Commission (FERC) was formed. FERC is the federal-level agency responsible for utility regulations, and it was a reorganization of the earlier Federal Energy Commission.

In 1996, FERC issued orders that encouraged market restructuring. In response, NEPOOL proposed that ISO-NE be formed as the independent system operator for New England. ISO-NE was created in 1997.

As we will see in later chapters, NEPOOL did not just quietly fade away when ISO-NE was formed. It kept many of its powers. You might say (if you are cynical) that the highly trained engineers and lawyers of ISO-NE act as staff to the Participants Committee of NEPOOL. The Participants are insiders—"stakeholders." Participant Committee members are owners of power plants, distribution utilities, and so forth. Basically, ISO-NE is staffed by technical experts, and NEPOOL consists of representatives of companies with skin in the game.

The Participants Committee is the most powerful committee in NEPOOL. It has six Sectors, described in its annual report and other documents.[45] The Sectors are Generation, Transmission, Supplier, Publicly Owned Entity, Alternative Resources, and End User. As a first approximation, each Sector has 1/6 of the vote. That means that each Sector has about 16.6% of the votes. Representatives of the Gas Industry (Gas Industry Participants) can attend meetings and Sector meetings but cannot vote.[46]

The voting rules are very arcane. The Alternate Resources Sector has somewhat fewer votes, and then there are Provisional Members who can vote as a "Group Seat" to up to 1% of the votes. To learn more than you ever wanted to know about the voting arrangements, you can refer to the NEPOOL agreement.[47]

Members of the Participants Committee pay an annual fee. Again, the rules are complex, but, basically, not-for-profit organizations and government agencies pay $500 a year for membership, while for-profit companies pay $5,000 a year.

How are the ratepayers represented in the Participants Committee? Not well, in my opinion. The End-Users Sector of the Participants Committee is 1/6 of the votes on the Committee. The NEPOOL list of End User Sector members[48] includes:

- Large electricity users such as Praxair, Inc.
- Government agencies such as the Massachusetts Attorney General's Office
- Advocacy groups such as the Union of Concerned Scientists

As a ratepayer, I do not have the feeling that the End-User members necessarily have my best interests at heart. Even if I felt the End-User Sector spoke for the ratepayers, they would still be only 16% of the Participants Committee vote.

A person can learn about NEPOOL by reading the bylaws and the sector rosters and the minutes of previous meetings. In other words, you can learn about NEPOOL by reading what NEPOOL chooses to publish. The NEPOOL meetings themselves are closed to the public. As it says on the Participants Committee page of the NEPOOL website,[49] guest attendance at the Participants Committee meetings must be cleared with the Committee chair. No statements made in NEPOOL meetings can be quoted or published to the public (except, of course, for what NEPOOL itself publishes).

Among other things, this means that no reporters can attend the meetings as reporters. The trade journal *RTO Insider* attempted to get a reporter to join in the End User Sector. This apparently caused a fair amount of excitement and fear within NEPOOL, and NEPOOL filed a request with FERC. NEPOOL wanted FERC's approval for a new rule that would ban press members from joining the Participants Committee.[50]

FERC ruled that such a ban would be discriminatory to reporters.[51] However, while reporters could now join the Participants Committee, the FERC ruling left the actual ban on reporting in place. In a related docket, *RTO Insider* filed a complaint, alleging that the closed-door policy violates federal law.[52] As of this writing, FERC has not ruled on that docket. Don Kreis, the Consumer Advocate of New Hampshire's Office of the Consumer Advocate, has written a clear opinion piece arguing against the FERC decision: "NEPOOL wins, transparency and electric ratepayers lose, at the FERC."[53]

Just to complete this narrative, I need to include information about how the Boards of Directors of ISO-NE and NEPOOL are chosen. Even if there is no reporter at the meetings, one can hope that Boards of Directors can provide some outsider perspective on what the staff of the organization is doing.

NEPOOL does not have a board of directors. NEPOOL officers are chosen from within the Participants Committee. The webpage lists its officers.[54]

In contrast to NEPOOL, ISO-NE clearly displays its board of directors on its webpage.[55] A quote from the webpage: "Board members have no financial interest in any company participating in New England's wholesale electricity markets." Many of the board members used to work for utilities or regulatory agencies and are now consultants. In my opinion, this is a correct mix. To be on the board of an agency such as ISO-NE, people need to have some expertise in the area in which the agency works. However, the board members should not be financially tied to the companies that are regulated. So far so good, about the ISO-NE board.

Next, so far so bad: "Candidates [for the ISO-NE board] also receive the endorsement of the NEPOOL Participants Committee." The NEPOOL "stakeholders" are still holding their stakes and their veto power.

The next chapters, on Jump Ball Filings, will show NEPOOL and ISO-NE in action.

Observing their filings, it is clear to me that the well-being of the majority of NEPOOL members is most relevant to NEPOOL actions. The well-being of end-users is less relevant. Now, onward to the Jump Balls.

CHAPTER 17

FIRST JUMP BALL

The rulings on winter reliability

WE WILL LOOK AGAIN at the challenge of how to keep the electricity flowing during a cold winter night in New England, as we did with the chapters on Winter Lights (chapters 8 through 10). This time, we will look at it from the point of view of how the regulations got put in place.

We will start with the final FERC ruling on the final Winter Reliability Program, several years before Pay for Performance was due to begin. In this example, we will see many of the familiar players, but playing in a situation that is far less complicated than Pay for Performance will become.

Winter Reliability Programs are simply about paying for fuel to be on-site. The only questions are about what types of fuel get payments and how much. The FERC ruling on winter-reliability planning in

New England was issued on September 11, 2015, and covered the winters of 2015–2018.[56]

As I described in chapter 4, the Angelic Miracle of the Grid, before there were RTOs, local utilities cooperated in regional power pools, which still exist. In particular, NEPOOL (the New England power pool) is still a very active organization.

NEPOOL did not give up all its powers and just go gentle into that good night simply because ISO-NE was created. Specifically, if the Participants Committee of NEPOOL does not approve of an ISO-NE Market Rule proposal, NEPOOL can propose an alternative Market Rule. NEPOOL can present its own Rule to FERC at the same time as the ISO-NE rule is presented. Two simultaneous filings on one topic is called a "Jump Ball" filing, after the method which is used to begin play in basketball. The FERC winter-reliability ruling (referred to above) gives a brief description of Jump Ball filings.

In basketball jump ball, an official throws a ball up into the air, and two opposing players attempt to gain control of the ball while it is in the air between them. In a NEPOOL-ISO-NE jump ball, two lengthy filings are submitted to FERC simultaneously by the two parties, and FERC makes the choice of which to implement. Or, FERC can combine the two filings into one rule, therefore making both parties unhappy. In this ruling for winter reliability, FERC chose the NEPOOL proposal.

So, now we have three sets of actors in this drama. There's ISO-NE with its proposal, and there's NEPOOL with its own proposal, and there's FERC, who will make the decision.

Jump-ball filings are quite common. I generally take examples from my own RTO, but similar types of filings happen at other RTOs. For example, the RTO that includes Pennsylvania, New Jersey, Maryland, and neighboring areas (PJM) filed two capacity-repricing proposals with FERC. In other words, PJM filed a Jump Ball with itself. They filed the first proposal because they thought it was a good thing, and they filed an alternate proposal because they got so much pushback from the states on their original proposal.[57]

Getting back to the first New England Jump Ball, we see that ISO-NE proposed a Winter Reliability Program that included all types of plants that can keep fuel on-site. In contrast, in its filing, NEPOOL wanted to limit payments to only those plants that could keep oil or LNG available, and to demand response. The ISO-NE filing treats coal and nuclear plants just like other plants that can keep fuel on-site and rewards them. The NEPOOL filing is basically about dual-fired gas plants. (Spoiler alert: FERC picked the NEPOOL version.)

The Participants Committee in NEPOOL is a mix of what are called "stakeholders." ISO-NE is more of a technical and regulatory group. According to page 9 of the FERC ruling referenced above, NEPOOL asserted that its proposal is supported by 87% of the Participants Committee, while the ISO-NE proposal was supported by only 13%.

ISO-NE and NEPOOL are not the only entities involved in the filing. There were also intervenors. These are organizations that are given status to intervene and comment upon the filing, and to try to affect the FERC order. In this ruling, everyone and their cousin was an intervenor. Many of the intervenors were also part of the Participants Committee.

The groups supporting the ISO-NE proposal were in favor of the fact that it included a more diverse group of energy sources, including nuclear and hydro, that have fuel on-site without any extra effort. The gas-fired generators and their various associations supported the NEPOOL proposal.

I do not know the details of the voting. There are no sunshine laws about the grid. When I grew up in Chicago, "insiders" making policy in secret was called "back-room dealings." On the RTO grids, such dealings are the standard process.

Of course, NEPOOL participants may well have proprietary data to protect. Still, it seems to me that their statements in evaluating a proposed ruling could be published publicly.

FERC chooses

IN THE WINTER-RELIABILITY ruling, FERC chose the NEPOOL version (pages 19–20 of the ruling). They admitted that ISO-NE was working with stakeholders to expand the types of resources eligible for payment, but that paying those resources would not be an incentive for the other types of power plants to keep fuel on-site for the winter. It would not be an incentive because the other types of power plants *already* keep fuel on-site for the winter.

In other words, FERC is fuel-neutral, paying all types of power plants the same money for the same results. Except when FERC isn't fuel-neutral.

Well, that is an oversimplification. In its ruling, FERC justified the reason that it was not fuel-neutral. It acknowledged that the entire Winter Reliability Program was "out-of-market." Therefore, according to FERC, the program does not have to be non-discriminatory as long as it does the job of keeping the lights on. In other words, we aim at having fuel-neutral markets, but once we do things differently and we are out-of-market, then pretty much anything we do is okay.

The final words of the FERC ruling (ruled in 2015) are a bit shocking to me, but I am not a lawyer. Here's a quote:

> We also disagree with commenters who argue that NEPOOL's proposal is unduly discriminatory and represents a collateral attack on the Commission's prior orders. As the Commission explained in the September 9, 2014 Order accepting the 2014–2015 Winter Reliability Program, also applicable here, the proposal is not unduly discriminatory merely because

> it does not compensate all resources for providing firm fuel service if those resources are not similarly situated. On rehearing of the January 20, 2015 Order, the Commission stated that "if any future out-of-market program is not fuel neutral, we expect that ISO-NE would provide a detailed description of the options it considered to make the program fuel-neutral and why those options were ultimately not included." ISO-NE was not obligated to expand the program. Rather, the Commission intended to encourage ISO-NE to work to expand the program, while still affording ISO-NE and regional stakeholders a reasonable amount of discretion to design a program that would adequately address the region's needs.

Did these guys ever watch *Star Wars*? In the words of Yoda: "Do or do not. There is no 'try.'"

In contrast, FERC's 2015 statement was the equivalent of: "Not obligated to do. Really, trying is all that is required." You don't have to be fair. What you do is not "unduly discriminatory" if you can show you tried.

In my opinion, this is a heck of a stupid way to run a grid.

CHAPTER 18

SECOND JUMP BALL AND PAY FOR PERFORMANCE

Pay for Performance

In the last chapter, we looked at the politics and concerns around the Winter Reliability Program for New England. We watched ISO-NE and NEPOOL file rival proposals to FERC (jump-ball filing) and watched FERC take NEPOOL's version, while admitting that it discriminated against many types of power plants. (But we tried!)

However, as I noted in the introductory chapters about keeping the lights on in winter, FERC insisted on a market-based solution to take the place of the Winter Reliability Program, starting in 2018. You will not be surprised to learn that both NEPOOL and ISO-NE once again filed schemes with FERC. Yes, there was another jump-ball filing.

Let's look at how this one worked out.

ISO-NE and NEPOOL filed their proposals with FERC simultaneously in a jump-ball filing.[58]

ISO-NE's filing begins right after the main transmittal letter, and the first ISO-NE document within that filing is Attachment I-1a. That document begins by describing how the current forward-capacity market is seriously flawed, to the point of actually being broken. Plants bid into the capacity market, giving the impression that plenty of capacity is available, but then, when those plants are called during a winter emergency, they say something like "Sorry, the cat ate my homework." Actually, what they say is "Sorry, I can't get any gas to run my plant."

Attachment I-1a explains how the current situation does not lead to reliability. The costs and rewards are economically perverse. On page 13, it states:

> Even worse than its effects on the investment decisions for individual resources, however, is the effect of this exemption-laden, flawed availability-based performance metric on the New England fleet as a whole. Because resources that do not contribute to system reliability during scarcity conditions earn the same capacity payments as resources that do, it is profitable for resources with low costs and poor performance during scarcity conditions to remain in the capacity market. These low-cost, but poorly performing resources displace higher-cost but better-performing resources. These higher-cost resources, because they would contribute more to system reliability, are actually more cost-effective than the resources that displace them. In effect, then, the current FCM (forward capacity market) has a structural bias to select less-reliable resources, an outcome completely at odds with the goals of maintaining reliability in a cost-effective manner.

A few pages later, page 16, is a section titled "The Penalty Rate in the Current FCM Rules Is Needlessly Complex and Is Too Low to Be

Effective." This section notes that penalty rates for non-performance actually decrease as the scarcity condition lengthens. As ISO-NE notes, "This perverse property is difficult to reconcile with economic logic."

ISO-NE has to run the grid, and their opinion of non-performing plants is pretty clear. However, despite ISO-NE's best efforts, the grid tends to become less reliable and more fragile.

The ISO-NE filing: the proposal

IN ITS FILING, ISO-NE proposed a new system, a two-settlement system. The core of the two-settlement system is that ISO-NE would run a capacity auction (as they do now) but would also designate "scarcity conditions" on the grid. The rules would be different during scarcity conditions.

In the first settlement, a power plant takes on a forward-capacity-supply obligation (as it does now) and gets a capacity-availability payment. The second settlement depends on what happens during a scarcity condition on the grid. If the plant contributes electricity up to its capacity-supply obligation during the scarcity condition, its capacity payment is not changed. If it contributes less during the scarcity, it loses capacity payments. The capacity payments that the non-performing plant loses are directed to plants that contribute more than their supply obligations during the scarcity condition. Also, a plant that didn't even bid into the forward-capacity market can get a capacity payment during a period of grid scarcity, just as if it were a plant contributing more than its capacity-supply obligations.

The description of the complex process for the second settlement process[59] contains a significant feature of Pay for Performance:[60] "A significant advantage of Pay for Performance is that it is resource-neutral. The same payment provisions apply regardless of resource type. This is in sharp contrast to the current FCM rules, which include separate monthly capacity-payment provisions for the various resource types."

The NEPOOL proposal

As you might expect, the NEPOOL proposal[61] is much nicer to non-performing plants.

The NEPOOL proposal invents the completely new metric "Equivalent Peak Period Forced Outage Rate (EFORp)." Payments to a power plant would not depend on how it performed during actual conditions of scarcity on the grid, but rather on how it improved over the years.[62] As often with NEPOOL filings, this proposal was sure to warm the hearts of the majority of the Participants Committee members.

The Pay for Performance ruling

FERC's basic ruling was in favor of ISO, though it included some aspects of the NEPOOL proposal.[63]

As stated on pages 9 and 10 of the ruling:

> First, NEPOOL's proposed EFORp metric is flawed. By measuring a resource's performance against only its own historical performance, NEPOOL's proposed EFORp metric may inappropriately reward poorly performing resources and penalize highly performing resources, which could further erode reliability in the region.

FERC also points out that power plants would have an incentive to degrade their own performance for a couple of years, in order to get "most improved" financial rewards in later years.

To convince yourself that the RTOs are not a market, I suggest you might want to go the FERC ruling and read the whole forty pages. You will see an example of the incredible complexity of the grid rules. Do the ISO-NE or NEPOOL proposals treat renewables right? Do they treat imported power right? How do they assess rewards for plants coming online quickly (10-minute and 30-minute reserves)? Are the performance penalties too high? Are the stop-loss provisions that hedge these penalties too high or too low? How will the External Market Monitor treat all these provisions? Is some group about to obtain "market power"? Near-monopoly power, market power, is considered a very bad thing.

Does Pay for Performance work?

SOMETIMES I WONDER WHETHER any of the people who derive these complex formulas "check their work," as we were all taught to do in grade school. ISO-NE proposed something, and FERC took part of the ISO proposal and part of the NEPOOL proposal and built a kludge system. As far as I can tell, nobody checked their work. On Labor Day 2018, Pay for Performance made an $8 million error over the course of one day.

It happened this way. On Labor Day, it was hotter and more humid than expected by ISO, plus Mystic Station went offline. The grid was in trouble. Prices were soaring to $2,600 per MWh ($2.60 per kWh).[64]

That much loss of generation capacity could have led to real problems on the grid itself, not just with the prices. However, the Balancing Authority dispatchers are great (the angelic miracle of the grid), and all was well. I have no idea whether Pay for Performance provided any incentive for plants to be online at that time.

Next came the Pay for Performance settlement, with capacity-payment penalties assessed to the underperformers and those penalties paid to overperformers — according to the formula.

The formula failed. ISO paid $7.8 million more to overperformers than it received from underperformers. To make up the difference, it had to assess a fee to every group who gets a capacity payment. There was a certain amount of consternation about these unexpected levies, even on overperforming plants. Also, some Energy Efficiency groups get capacity payments but were not eligible to participate in this particular transaction because it happened in off-peak hours. (I told you it was a kludge of a system.) The Energy Efficiency groups have to pay "their" part of the $8 million, even though, technically, they were not eligible for over-or-under penalties or payments at that time.[65]

I suspect that there will be a further patch, a longer formula, to attempt to fix this type of problem. With nobody responsible for the grid, the unexpected consequences never seem to end.

Let's look briefly at a third-party analysis of Pay for Performance. ICF International, a technology-services-consulting business, wrote a white paper on Pay for Performance. It explains the penalty/reward formula and the implications of the formula.[66]

ICF International expects that Pay for Performance will lead to "modernization of the fleet," which will happen due to the retirement of "inflexible ... steam-fired units." They are hopeful that the retirements will be orderly and not lead to too many price spikes.

In other words, the ICF evaluation predicts more gas online and the retirement of steam-fired units (coal and nuclear), which keep fuel on-site.

Hey, this is where the problem started! The more gas online, the more issues about getting gas in winter, and therefore, the more Pay for Performance will be needed and perhaps expanded. (As if penalties could help plants get gas when it is not available.) If you want to build a gas plant, Pay for Performance is your kind of regulation.

ICF International also predicts that plants will be more uncertain about their capacity payments.

Of course, the point of Pay for Performance is to increase this uncertainty: if your plant doesn't go online, you won't get paid. On the other hand, much of the uncertainty is totally out of the control of the power plants. How many hours will the grid be in a stressed situation that triggers Pay for Performance? How stressed will the grid be? Will your plant get online and get paid, or not get online and have to pay others?

The original point of capacity payments was to provide a backstop for plants that follow load and don't run all the time. Apparently, this worked too well: plants got the capacity money, but the grid still didn't have all the resources it needs. Pay for Performance erodes part of the backstop, in the hope that this will lead to more reliability on the grid. Maybe it will (I doubt it), but maybe we will need a couple more rounds of regulation to fix the problems created by Pay for Performance.

Unlike questions that concern renewables or greenhouse-gas initiatives, most people agree that plants should be ready to run on a cold day. Yet each individual plant is owned by a different entity (some of the entities owning more than one plant, of course), and nobody is in charge of the grid.

This makes the situation virtually impossible to manage. Nobody can just say, "These twelve plants are our designated cold-weather plants, and we will pay them to keep fuel on-site." In an RTO, that would not be fair to all the other plants. In an RTO, there clearly needed to be a new type of auction or something.

The RTO must strive for "fairness" and "to keep the grid functioning." Somehow, for "fairness," being "fuel-neutral" has become a major obligation of the RTOs. How to get plants to keep fuel on-site without actually *saying*, "Keep fuel on site"? It is a difficult problem but provides much employment for grid experts and lawyers.

I will be blunt here. FERC is supposed to be fuel-neutral. Any special support for low-emissions plants is supposed to come from the states, from other federal regulators (such as the EPA) and so forth. And yet, somehow, when it comes right down to it, for the winter-reliability projects of the last chapter (First Jump Ball), FERC had no difficulty preferring the NEPOOL over the ISO-NE plan. Because the NEPOOL plan does not reward nuclear, coal, or hydro for being available when the grid is stressed, the ruling was another de facto vote in favor of gas-fired plants.

After the FERC ruling that led to Pay for Performance (Second Jump Ball), a third-party analysis concluded that Pay for Performance would encourage the retirement of older, steam-fired plants. This would be partially due to gas-plant payments for coming online quickly, compared to the financially neutral situation of a steam plant being steadily available.

I didn't cover the speed-of-coming-online part of the ruling in this chapter: sometimes I think I ought to write a book called "Jump Ball, the Complete Story." But this is not that book. I do cover speed of coming online, to some extent, in a January 2017 blog post.[67]

Once again, the FERC ruling on Pay for Performance favors gas-fired plants.

For FERC, Gas is Great.

These rulings (and others) in the RTO areas require immense amount of time from many lawyers. And the rulings still manage to support gas-fired plants above all else. I think this sort of thing is what led Department of Energy Secretary Perry to think that something had better be done about the fuel-security situation on the grid.

Perry is from Texas, and I think he probably has no particular problem with having lots of natural gas on the grid. "No gas!" is not likely to be his issue. Much of the Permian Basin (a major oil and gas resource) is in Texas. However, natural gas is just-in-time delivery. Perry was trying to ensure the security of the grid in bad weather, by rewarding plants that keep fuel on-site.

Perry may have noticed that FERC was happy with complex formulas to keep oil on-site at a gas plant but was unwilling to reward plants that kept fuel on-site as a natural course of business. That is FERC's view of "fuel neutral." Perry's approach was heavy-handed, and it failed. With Pay for Performance, FERC once again ruled on fuel neutrality as "neutral in favor of gas-fired plants."

In my opinion, something needs to be done about fuel security for the grid. The RTO grids are moving toward fragility and rolling blackouts. Personally, I don't think that the RTOs are capable of solving this problem.

But let's look forward. Our next step is the Fuel Security Study by ISO-NE.[68]

Late-breaking news. On June 18, 2020, FERC ruled in favor of an "Inventoried Energy Program" which compensates power plants for "inventoried energy" that is kept on site at a generating facility. This ruling includes payments for plants that routinely keep fuel inventories on site. Payments can be made to oil, coal, nuclear, hydro and biomass plants. In other words, FERC enacted a type of Winter Reliability payments for the winters of 2023 and 2024, in which the stored fuel is called "inventoried energy." The Inventoried Energy Program is far more fuel-neutral than the earlier Winter Reliability Program described in chapter 17. That program mostly paid for dual-fuel gas-fired plants to keep oil onsite.

Not everyone liked the fuel-neutrality of the Inventoried Energy Program. FERC Commissioner Richard Glick wrote a dissent in which he said that, for fuel security, it would be more effective to burn the money than to pay plants that already keep fuel onsite. I disagree. In my opinion, this new program is a tiny step toward encouraging a resilient grid that does not totally depend on just-in-time gas delivery.

Unfortunately, the new Inventoried program is not a substitute for the complex formulas of the Pay for Performance program, described in this chapter. Pay for Performance and Inventoried Energy will run simultaneously. The new program is just another level of bandaid for the winter grid in New England. The FERC docket for the new program is ER19-1428-003.

CHAPTER 19

THE ISO-NE FUEL SECURITY REPORT

The past is prologue

IN EARLIER CHAPTERS on Winter Lights (chapters 8–10), we reviewed the ISO-NE grid during the frigid days around New Year's Day, 2018. We saw how oil saved the grid. We had drama (icebreakers to get the fuel ships through), more drama (oil inventories sinking at plants that could burn oil, and horrible weather slowing deliveries) and even more drama (some of the LNG delivered to Massachusetts was sourced in Russia). The ISO-NE operators made many emergency conference calls.

Actually, it is not fair to call this mere "drama," though it was dramatic. There was a real problem. The grid was teetering on the edge of rolling blackouts.

Was all this excitement an aberration during a cold snap or a warning for the future?

As you can guess, it was a warning.

As a follow-on to Secretary Perry's ill-fated attempt to increase fuel security on the grid, the Department of Energy asked all the RTOs to study their grid's vulnerability to fuel-security problems. At the time of the mid-winter issues on the grid in late 2017, ISO-NE was almost ready to publish its fuel-security study. The crisis on the grid ended on January 8, 2018, as the weather warmed up. The study, "Operational Fuel-Security Analysis," was published on January 17, 2018.[69]

ISO-NE initially ran 23 scenarios for the future of the grid. In 19 of the scenarios, the grid would have rolling blackouts by the winter of 2026.

There was only one scenario where the grid had a solid, no-emergency operation. That no-problem scenario had very optimistic assumptions and included increased deliveries of LNG (liquified natural gas).

The future comes closer

In the fuel-security study, ISO-NE modeled the temporary closure of a major gas-fired plant, Mystic Station. This temporary closure would lead to rolling blackouts. The time-frame that ISO-NE modeled was the winter of 2025–2026.

In March 2018, Exelon, Mystic's owner, announced it planned to close the plant permanently in 2022.[70]

I wrote about this, of course. On May 19, 2018, the *Valley News* printed my opinion piece "Rolling blackouts are probably coming to New England sooner than expected."[71]

The section below is an edited version of my article.

****

When there's not enough supply of electricity to meet demand, a grid operator cuts power to one section of the grid to keep the rest of the grid from failing. After a while, the operator restores the power to the blacked-out area and moves the blackout to another section. That is a "rolling blackout."

The New England grid operator, known as ISO-NE, recently completed a major study of various scenarios for the near-term future (2024–2025) of the grid, including the possibilities of rolling blackouts.

In New England, blackouts are expected to occur during the coldest weather, because that is when the grid is most stressed. Rolling blackouts add painful uncertainty — and danger — to everyday life. You aren't likely to know when a blackout will happen, because most grid operators have a policy that announcing a blackout would attract crime to the area.

In early April, Chicago-based energy provider Exelon Corp. said it would close two large natural-gas units at Mystic Station, Massachusetts. In its report about possibilities for the winter of 2024–25, ISO-NE had included the loss of these two plants in one of its scenarios. The ISO-NE report concluded that Mystic's possible closure would cause 20 to 50 hours of "load shedding" (meaning rolling blackouts) and hundreds of hours of grid operation under emergency protocols.

When Exelon made its closure announcement, ISO-NE realized that the danger of rolling blackouts was suddenly more immediate than 2024. It now hopes to provide "out-of-market-cost recovery" — subsidies — to persuade Exelon to keep the Mystic plants operating. If ISO-NE gets permission for the subsidies from the Federal Energy Regulatory Commission, some of the threat of blackouts will retreat a few years into the future.

Ominously, 19 of the 23 ISO-NE scenarios led to rolling blackouts. The one "no-problem" scenario (no load shedding, no emergency procedures) is one where everything goes right. It assumed no major pipeline or power-plant outages. It included a large renewable build-out plus greatly increased LNG delivery, despite difficult winter weather.

This positive scenario is dependent on increased LNG deliveries from abroad. That's because the Jones Act, a section of the Merchant Marine Act of 1920, prohibits ships built and registered outside the U.S. from delivering goods between American ports. There are no LNG carriers flying an American flag, so New England cannot obtain domestic LNG. It must import foreign LNG, which can be delivered by foreign-flag ships.

We could plan to import more electricity from Canada, instead of importing more fuel, but ISO-NE notes that such imports are problematic. Canada has extreme winter weather (and curtails electricity exports) at the same time that New England has extreme weather and a stressed grid.

To avoid blackouts, we need to diversify our energy supply beyond renewables and natural gas to have a grid that can reliably deliver power in all sorts of weather. We need to keep existing nuclear, hydro, coal, and oil plants available to meet peak demands, even if it takes subsidies. Coal is a problem fuel, but running a coal plant for a comparatively short time in bad weather is a better choice than rolling blackouts.

This can't happen overnight. It has to be planned for. If we don't diversify our electricity supply, we will have to get used to enduring rolling blackouts.

Alarmist propaganda and pollyanna scenarios

My op-ed appeared on May 19, and Don Kreis wrote a reply. His letter to the editor appeared on May 24 in the same newspaper. Kreis serves as consumer advocate in the New Hampshire Office of the Consumer Advocate. He wrote:[72]

> But it is hard to blame Angwin for succumbing to the alarmist propaganda emitted by regional grid operator ISO New England, which should know better.

Thinking of the people and the presentations that I see quite regularly at ISO-NE meetings, "alarmist propaganda" seems a bit much. In some of the meetings, staying awake was the real challenge. I would have welcomed some alarmist propaganda.

But let's leave rhetoric aside for a bit. No, I can't do it. I gotta repeat one little rhetorical statement, made by a friend of mine:

> ISO-NE has the responsibility for keeping the lights on. I don't think CLF has that responsibility, right? So, CLF doesn't have to be alarmed at anything. Someone else is going to take care of it for them.

Okay. I have now mentioned CLF, the Conservation Law Foundation. After the ISO-NE fuel security report appeared, the New Hampshire Office of the Consumer Advocate and CLF, plus other groups, sponsored a rival fuel-security study, which I will call the "Synapse Report."[73] This rival study was written by Synapse Energy Economics, and Kreis mentioned the study in his letter to the editor.

The Synapse Report painted a much rosier picture of the grid and claimed that ISO-NE had made overly conservative assumptions.

Indeed, the minor exchange between Kreis and me was just the tip of an iceberg. The base of the iceberg was a bigger dispute. Once again, it was between an ISO-NE analysis and part of the Participants Committee.

Before we go further, here's a timeline for this controversy:

- **January 18, 2018:** ISO-NE issued its fuel-security study with 23 scenarios. It predicted rolling blackouts in 19 of the scenarios.

- **January 26:** ISO-NE held a meeting with the NEPOOL Participants Committee to present its report and ask for comments.
- **February 15:** comments were due. Many members of the Participants Committee submitted comments, and ISO-NE ran more than a hundred more scenarios.
- **March 29:** Exelon announced its plans to close Mystic Station.
- **April 26:** ISO-NE issued a deck (set of viewgraphs) with results from analyzing the hundred new scenarios.[74] Pages 44 to 61 of this deck, which I will call the "Addendum Report," are the results of the sixteen additional scenarios requested by a group called the "Joint Requesters."
- **May 3:** Synapse Energy Economics issued its report "Understanding ISO New England's Operational Fuel Security Analysis." This report is mainly based on the ISO-NE analysis of the Joint Requesters sixteen scenarios, with some discussion of other requested scenarios.
- **May 19:** my column appeared in the *Valley News.*
- **May 24:** Don Kreis's letter appeared in the *Valley News.*

Note: What do I mean by "deck"? ISO-NE does not always issue a formal report on a topic but instead posts a PDF of the viewgraphs used for a presentation. This posting is generally referred to as a "deck," that is, a viewgraph deck.

Note: RENEW Northeast[75] is a nonprofit association with the goal of promoting renewable energy. RENEW Northeast is not itself a member of the Participants Committee, but many of its members are members of that committee. RENEW Northeast is listed as a member of the Joint Requesters and as a sponsor of the Synapse Report.

As I hope I have shown through this timeline, this controversy is about ISO-NE and the Participants Committee. I think another Jump Ball will be in the future.

The pollyanna scenarios of the Addendum Report

In the Addendum Report, we can see that the Joint Requesters requested a base-case scenario with:

- more LNG available (1.25 Bcf per day base case instead of 1.0 per day in the ISO-NE base scenario)
- 1000 MW more electricity imported from Canada (3500 MW instead of 2500 in the ISO-NE base)
- Renewables at 7900 MW instead of 6600 MW
- Slower growth in gas use for homes

The members of the Joint Requesters sponsored the Synapse Report, and it is based on their scenarios. In the next chapter, I will review the Synapse Report. It is shorter than the Addendum Report, and it is a readable report, not a "deck." In my opinion, the Synapse Report shows how the Participants Committee is going to act in the future.

The Synapse Report has cheerful, reassuring conclusions about the grid. It's a shame they are incorrect.

CHAPTER 20

POLLYANNA SCENARIOS ON THE GRID

Everything is rosy with misplaced concreteness

ONE OF THE WAYS that the Synapse Report achieves happy outcomes is with misplaced concreteness. There's a name for taking-targets-as-actual-reality. In logic, it is called the "fallacy of misplaced concreteness" or "reification." A more common description for this fallacy is sometimes stated as "the map is not the territory." We will see this fallacy in action in several sections of the Synapse Report.

Misplaced concreteness for renewables

The Synapse Report claims the original ISO-NE report underestimates the amount of renewables that will be available to the grid. Synapse notes that state clean-energy laws require much more renewable energy than ISO-NE uses in its predictions.

In other words, despite the fact that states and even countries (see Germany) routinely fail to meet their renewable-energy targets, Synapse claims that ISO-NE should take those targets as the reality for planning.

Low-probability events

ISO-NE looked at the consequences of certain types of events, such as a compressor failure on a major gas pipeline. Such events would often lead to rolling blackouts. Synapse said that ISO-NE should have looked at the low probability that such events would take place. Perhaps the idea behind this Synapse objection was that these events are so unlikely that the ISO-NE report was like modeling a meteor strike: not worth the time it takes for the modeling. If it happens, it happens. Nobody can prepare. (This would not be a good attitude for a grid operator.)

Synapse's "It's a rare event — so don't worry so much" idea was proven false within weeks of the ISO-NE report being issued. The report predicted that if Mystic Station were offline for an extended period during the winter, this would lead to around 20–50 hours of rolling blackouts. Shortly after the ISO report was issued, owners of the Mystic Station announced their plans to take the plant offline permanently.

Quick fixes are available

A third Synapse idea was that, if these failures or shutdowns happened, workarounds would be managed, so ISO-NE assuming a lengthy outage or failure was alarmist. However, the plans for closing Mystic certainly shows that quick fixes are not guaranteed and that outages can be lengthy or permanent.

Everything is rosy with electricity imports

SYNAPSE IS VERY UPBEAT on how much electricity will be imported by 2024.

Electricity imports in general

Enjoying another excursion into the fallacy of misplaced concreteness, Synapse castigated ISO-NE for not acknowledging that a transmission line carrying 1000 MW of power from Canada was scheduled to be built. As a matter of fact, Synapse castigated ISO-NE as follows: "The ISO believes that uncertainty about the construction of a transmission line to Canada disqualifies the additional imports for use in its Reference case" (page 4 of the Synapse Report).

"Scheduled" is too strong a word for the status of that power line. Conservation Law Foundation supports the line, and the line has many of its permits, but my own experience is that such lines are problematic to build. For one thing, they run through one of the two north-south states (New Hampshire and Vermont), but they don't deliver any power to those states. The power goes to Massachusetts.

This means that people in the north-south states have very little reason to welcome the lines. Consequently, people in Vermont and New Hampshire are likely to protest the lines, lie down in front of the bulldozers, and so forth. Sometimes (many times) such lines simply don't get built around here. ISO-NE believes that the uncertainty of construction of that power line is an important factor, and I believe it, too.

Electricity imports in winter

In assuming we can get more electricity from Canada (or even the same amount as we obtain now), the Synapse Report doesn't look at important evidence: the history and the contracts of imports from Canada in the winter. ISO-NE was modeling what happens in cold weather, because that is when New England is most likely to have power shortfalls. Canada has really cold weather at the same time New England does. Canadians are aware of this. Therefore, there have been and will be shortfalls in cold-weather power inputs from Canada.

As a matter of fact, the original ISO-NE report was too rosy on imports from Canada. Even ISO-NE assumes we get twice the amount of energy that we are likely to get from Canada in cold weather.

The original ISO-NE report (pages 51–52) describes how the Canadians do not enter the New England Forward Capacity Market at the level that Americans buy their electricity. A resource that enters the Forward Capacity Market gets a "capacity payment" and takes on a "supply obligation."

ISO-NE analysis

ISO-NE notes in its analysis:

> Each (ISO-NE fuel security) scenario assumes a level of imports (from Canada) at least twice the amount obligated through the Forward Capacity Market. In other words, half the imports assumed in the study may not be available if the neighboring area where they are located needs them. This is important because Québec, New York, and New Brunswick all experience winter weather at the same time as New England.

At any time, Canada can ship New England only half the usual amount of electricity, without any penalty. No supply obligation, no foul.

In short, ISO-NE's supposedly "alarmist propaganda" assumes we always get at least twice the electricity that the Canadians are obligated to supply us. Not very alarmist. The ISO-NE predictions in this area are a bit too rosy, if you ask me.

Ignoring the history of imports

Before I was fully aware of the implications of the New England forward-capacity market, I was aware of Hydro-Québec and the way its electricity tended to disappear when New England needed it most. When the winter weather makes the going tough in New England, the Canadians get tough with New England. When there is really a cold snap, the Canadians ship less energy south, because they need the energy at home.

In January 2013, I wrote a blog post, "Cold Weather Winners and Losers on the Vermont Grid."[76] In that post, I noted that, during a spell of extremely cold weather, Canada shipped only 50% of the electricity to us than it usually shipped. I didn't know about the Canadian Capacity Supply Obligations when I wrote that blog post in 2013. But I could easily see that the Canadians didn't send us power when we really needed it.

In terms of electricity imports, the Synapse Report is beyond wishful thinking. It ignores history (such as how little electricity Canada supplies us during a cold snap). It ignores contracts and obligations about electricity sales. It ignores how small an amount Canada takes on as a Capacity Supply Obligation to New England.

Ignoring history and contracts pushes Synapse beyond wishful thinking and heads toward just plain "alternative facts" about electricity imports.

Everything rosy with LNG imports

LNG: Every day a peak. The Synapse Report castigated ISO-NE for assuming that LNG deliveries for a base case would be 1.0 Bcf (Billion Cubic Feet) per day. Now, ISO-NE also modeled higher and lower deliveries. Still, Synapse set its base case differently. Synapse set its base case at 1.25 Bcf per day.

To quote the Synapse Report (page 4):

> In other scenarios, the ISO report varied the daily-delivery quantity from 0.75 to 1.25 Bcf/day. *Note that actual imports in February 2016 totaled 1.25 Bcf/day.* [Emphasis added]

Hmm. Another alternative fact: the actual statement in the ISO report is on page 34:

> The maximum amount of re-gasified LNG that can be injected into New England's pipeline system is about 2.04 Bcf/d. In recent years, the most LNG injected at any one time into New England's pipeline system was 1.25 Bcf/d, *on one day* in February 2016. [Emphasis added]

Funny how a one-time peak day turns into the base amount in the Synapse Report.

As ISO points out repeatedly in its report, it takes days to order up new LNG, and there is competition for this gas going into northern pipelines (here and in Canada). There is also worldwide competition for LNG. But somehow, in Synapse, the maximum day is the base case.

Natural gas pipelines (not being built)

We are maxed out in New England. For once, ISO-NE, Synapse, and I all agree: no major new natural-gas pipelines have been built in New England in many years, and none are likely to be built in the near future. Both the ISO report and the Synapse Report assume that no major new gas pipelines are built. (Reports assume that some local pipelines are expanded.)

In several areas of New England, natural-gas companies are refusing to hook up any more homes.[77] Synapse assumes a very slow rate of increase in home use of natural gas.

For once and only once ... I agree with Synapse here. New England won't be using more natural gas in the future. If gas companies won't do more home hookups, there won't be a major increase in the home use of natural gas. What about manufacturing use of natural gas? With no new pipelines, manufacturing won't use more natural gas. New England will not be using more natural gas in the future, because we won't be able to obtain it.

A note about everywhere: This chapter is about New England and the Synapse report and so forth. However, the pipeline issue is a nationwide issue. When areas close their nuclear and coal plants, and move to natural gas, they often cannot build pipelines to handle the increased quantities of natural gas that they will need. Pipelines are lengthy things: they go through many areas and many jurisdictions. Since they carry fossil fuels, building a pipeline meets stiff resistance. No matter how much shale gas we find, it is the size of the pipeline that determines how many power plants can be supported. Inability to supply natural gas is yet another way that the grid becomes fragile.

Natural gas from Canada will still be rosy

Many wonderful things that Synapse feels would secure our grid are supposed to come from Canada in the future. Synapse expects that we will get increased electricity imports, but that won't work. First, winter happens in Canada at the same time it happens here. Second, we cannot assume that new transmission lines will be built.

But what about natural gas? The Synapse Report does not include any concerns that New England will probably receive less natural gas from Canada in the future. In contrast, in the main ISO-NE report, there are many such warnings. The first warning is about the depletion of natural-gas fields off the Maritime provinces, just north of New England. The report says: "Two natural-gas fields off the coast of Nova Scotia—Sable Island and Deep Panuke—are expected to be depleted before 2025."[78] As those fields get depleted, Canada will have less gas to export. Meanwhile, on high-demand days, the situation will get worse.

As the natural-gas fields become depleted, ISO-NE says, "(Depletion) would leave the Canadian Maritimes with just two sources of natural gas: Canaport and the Maritimes and Northeast (M&N) pipeline that carries gas between Canada and Maine." But would the Canadian gas get to Maine and to the Northeast? Basically, no, it wouldn't. As ISO-NE writes: "Under these conditions (high-demand days), the M&N pipeline will function as an internal distribution system, carrying gas from the west to Canada, rather than as a separate source of gas from Canada to New England."[79]

In other words, as the offshore fields get depleted, we can expect Canada to supply less natural gas to New England than it supplies now, especially in cold weather, when both Canada and New England have high demands for gas.

That doesn't sound good for fuel security in New England.

Take off the rosy glasses

In conclusion, thinking of Canada as an endless source of gas and electricity for New England is not a reasonable way to look at the situation.

Even the ISO-NE original studies make the rosy assumptions that Canada delivers twice as much electricity to New England as they are obligated to deliver.

The Synapse Report also looks at state policies. It makes this comment on the ISO-NE Fuel Security Report: "Two of the most unreasonable assumptions were the devaluing of state renewable portfolio standard (RPS) goals and the failure to account for existing Massachusetts law that requires imports of clean energy."[80]

Synapse assumes that all renewable targets will be met. After all, laws have been passed.

Meanwhile, with the new and rosier scenarios asked for by the NEPOOL members (addendum report) and the Synapse report sponsored by the Conservation Law Foundation and others, ISO-NE is pushed into a more-optimistic and less-realistic framework.

This is not good for New England or for planning but is a direct consequence of RTO governance.

Fuel-security studies and their implications

THIS ISSUE ILLUSTRATES the problems in the RTO areas, even without dealing with costs, renewables, and so forth. As soon as I begin talking about prices or the role of renewables, there's often a lot of virtue-signaling. People begin to claim that prices and reliability don't really matter so much compared to saving the planet. So, I wanted to just look at some other facts. Facts that are not so "fraught." I want to show that the RTO areas mishandle the grid, in even straightforward cases.

But the main reason I wanted to show these comparisons is—

We can't run a grid on wishful thinking. In the RTO areas, wishful thinking often prevails. The consequence of wishful thinking is a fragile grid.

Even if the RTO itself tries to be realistic (as ISO-NE has done in these reports), other organizations will attempt to undercut its conclusions. RTO areas are often more controlled by insiders than is good for governance. Insiders have strong financial reasons to see certain conclusions finalized. For them, it is not just wishful thinking. It's about money.

The insiders have a powerful platform for their views. They make up the power pools (such as NEPOOL). The power pool groups are generally not accountable to anyone. They don't even have a board of directors. Sometimes they are closed to the press. And they have strong standing before the regulator.

When upbeat speculation driven by insiders meets actual, real-world gas wells and constraints ... things don't always work out to be so upbeat for the grid. Some "stakeholders" deny these truths. Nevertheless, Reality Bats Last.

CHAPTER 21

THREE DAYS TO SECURITY

The future of fuel security

In earlier chapters (the Jump Balls), we watched ISO-NE attempt to provide steady power in winter. ISO-NE first arranged to buy actual fuel for dual-fired plants (the Winter Reliability Program). Then this program got slapped down by FERC as not being "fuel-neutral." FERC said that ISO-NE had to come up with a plan that is market-based and fuel-neutral.

Therefore, instead of the Winter Reliability Program, New England implemented the complex and probably unworkable Pay for Performance. This is supposedly fuel-neutral and market-based.

What is the more general future for fuel security in the ISO-NE area? Well, ISO-NE wanted to provide extra funding for the large Mystic Generating Station, the station that had announced its retirement, as described in chapter 19 on the ISO-NE Fuel Security report. In May 2018, ISO-NE asked FERC to be allowed to provide funding to the plant to ensure fuel security. I think this FERC order about various hearings is a good description of the interlocking dockets that came out of the ISO-NE request.[81]

FERC turned down the ISO-NE request. FERC said that ISO-NE had to make more general plans for future fuel security and present them by October 2019. And guess what? The new plans had to be market-based.

There's a certain déjà vu quality about this requirement. The déjà vu doesn't look any better than it looked before.

ISO-NE wrote an Energy Security Improvement discussion paper,[82] which they presented (yes, you know this is what would happen) at a NEPOOL meeting in April 2019. *RTO Insider* soon summarized the paper.[83]

NEPOOL rejected the ISO-NE proposal. However, ISO-NE went ahead and filed its proposal with FERC.

I think ISO-NE, which is a group of engineers and economists, understands the fuel-security problem very well. The Energy Security discussion paper has a clear description of perverse incentives that lead to the problems in the Northeast:

> To facilitate productive discussions of the ISO's concerns and potential solutions, this paper begins with a deeper examination of the underlying problems and their root causes. Our focus is whether the ISO-administered wholesale electricity markets—which were not originally designed for the challenges just-in-time generation technologies have wrought—provide adequate financial incentives for resource owners to make additional investments in supply arrangements that would be cost-effective and benefit the power system at times of heightened risk.

> Our central conclusion is that, in many situations, the answer is no.... The root cause is logical enough. Making these discrete investments, if they meaningfully reduce the risk of electricity-supply shortages (and therefore the risk of high prices), entails up-front costs to the generator—yet reduce the energy-market price the generator receives.... (This) results in a divergence between the social and private benefit of the investment—a situation we call a misaligned incentives problem.

To paraphrase: if a generator decides to store oil for days when the grid is stressed, the generator must spend money upfront to buy the oil. However, having the oil (especially if many generators decide to do this) will alleviate price spikes, and the generator will get less money when the price spikes because gas is constrained.

Keeping fuel on hand is a lose-lose game for the generator. Upfront expenditures will lower future revenues.

In the section "Oil Almost Fails at Saving the Grid," (part of chapter 9), I refer to the ISO-NE "Cold Weather Operations" report on the grid functioning during the December 2017–January 2018 extreme cold weather. That report shows a bar graph about which generators burned oil to keep the grid going.[84] Generators that were part of the Winter Reliability Program were the generators that supplied the grid with oil-based electricity when gas was not available. As the graph shows, Winter Reliability Program generators supplied about 95% percent of the oil-based electricity that was used during the cold snap.

In short, if ISO-NE paid for oil for winter reliability, the power plants were willing to buy and store oil. Without ISO-NE funding, they didn't buy oil. This is perfectly in line with the perverse incentives on the grid. The power plants do best when the grid is doing worst.

When many power plants have uncertain fuel supplies, and the grid is stressed, then electricity prices rise. At that point, many power plants make excellent profits. They don't even have to manipulate the markets to do this: the incentives are perverse enough on their own.

I don't know what will happen to secure our fuel supplies in the Northeast. As I write this, nothing has been resolved. There are studies, and there are proposals filed to FERC, but there is no resolution.

However, the proposals don't look good.

Once again, FERC has insisted on a "market-based solution," in which actual fuels are not named. Once again, ISO-NE is preparing an elaborate scheme to ensure fuel security without naming fuels or appearing prejudiced in favor of fuels that can be stored or appearing prejudiced against just-in-time fuels like wind and natural gas. It's "Pay for Performance, Round Two."

Michael Kuser at *RTO Insider* reported on a NEPOOL market committee meeting in June at which the fuel-security proposal was discussed.[85] At that time, several other groups (Massachusetts Attorney General's Office, NextEra, and others) also presented possible fuel-security schemes.

Three days to security

As you might expect, the ISO-NE proposal for fuel security consists of setting up more auctions. Their FERC proposal had to be market-based, and market-based turns out to mean new-kinds-of-auctions. ISO-NE was asked to solve a problem, but the request came with so many constraints that a solution is basically impossible. However, ISO-NE dived right in and came up with three new auctions.

The ISO-NE proposal is bewilderingly complex. It makes Pay for Performance look simple. Specifically, ISO-NE proposed three core components to improve the fuel-security problem. The three core components are described in their energy security discussion paper[86]:

> **Multi-day-ahead market.** Expand the current one-day-ahead market into a multi-day-ahead market ... with multi-day clearing prices for market participants' energy obligations.
>
> **New ancillary services in the day-ahead market.** Create several new, voluntary ancillary services ... (that compensate for) the flexibility of energy "on demand" to manage uncertainties each operating day.
>
> **Seasonal forward market.** Conduct a voluntary, competitive forward auction that provides asset owners with both the incentive and necessary compensation to invest in supplemental-supply arrangements for the coming winter.

Two of these new markets will be voluntary. As I see the proposal, only the multi-day-ahead market would be required. Okay, three new auctions, and I am understating the complexity of it all. As described in the *RTO Insider* article above,[87] there are actually other core components to the ISO proposal, including a forward rate, a spot rate, trigger conditions, and more.

The fundamental idea seems to be that, with more markets, including a market with multi-day clearing prices, generators will have enough incentive to keep fuel on-site.

Alternative proposals suggested by other participants at the NEPOOL meeting included sealed bids, new Reserve Products that can be purchased by ISO-NE in the day-ahead market, and ways for ISO-NE to be sure that gas-only resources have firm commitments for natural gas.

I can't possibly try to evaluate all the ideas, so I will look only at the ISO-NE proposal.

To even begin to describe the proposed changes to the day-ahead markets, I need to start by explaining the workings of the current day-ahead market. Even this market is quite complicated, and it tends to favor gas-fired plants financially. (You probably suspected that would be the case.)

The current day-ahead market

HERE'S HOW THE CURRENT day-ahead market works.

- In the first step, ISO-NE studies the weather projections, the history of electricity use on similar days, and any expected unusual events (such as the World Series) that can affect electricity demand.
- In the second step, ISO forecasters estimate the grid demand in five-minute intervals.
- In the third step, ISO asks power plants to bid into the day-ahead market in one-hour intervals, filling up the demand that it expects to meet.
- In the fourth step, ISO pays for the power it will buy the next day.

You read that right. ISO-NE pays ahead of time.

Power plants bid into the day-ahead market at their bid rate: for example, a gas plant knows how much it costs to make the next kWh of power, and that is what it bids in. (ISO-NE also knows what it costs to make the next kWh of power, and it checks that the bids are correct.) If the power-plant bid is low enough, the plant makes it into the list to supply power for that one-hour interval, and it gets paid the clearing price for that time interval. It receives the price bid by the highest bidder who was accepted for that interval. This was described in chapter 14 on "How Power Plants Get Paid."

You might say that this is a simple and straightforward situation, with perhaps adjustments at the end of the day if more or less power is needed. After the auction, ISO has enough power, and it knows it. ISO knows the lights won't go out; the power plants are paid around 24 hours in advance. What could be more simple and better? Finally, something that works right.

But no. The day-ahead market doesn't work right. The day-ahead-market payments are often higher than the real-time prices, and gas plants get to pocket the difference. Payments in the day-ahead market do not control actual power-plant dispatch.

Let me repeat this:

Payments in the day-ahead market do not control actual power-plant dispatch.

Enter the real-time market

A PLANT CAN BID INTO the day-ahead market, get paid, and then not get dispatched to provide power. There are some obvious situations that would lead to this, for example, if ISO-NE overestimated the need for power, and not all the plants that bid into the day-ahead are actually needed in real time. But it is more complex than that.

Let us say that a gas-fired plant has bid in the day-ahead market to supply power at 10 cents per kWh in the interval from noon to 1 p.m. It gets paid at the clearing price for the power it will supply in that interval. Let's say that the clearing price is also 10 cents per kWh. However, the next day, the wind springs up just before noon. Now we are in the real-time market. Dispatch is by cost. Like the day-ahead market, we have economic dispatch—lowest-price plants dispatched first. However, the real-time market is a physical system, with plants and wires, so dispatch is also by physical location and other constraints. "This plant bid into the day-ahead market" is not a constraint.

So, the wind is blowing in real time, and here comes a wind turbine. It gets outside funding by RECs and production tax credits. The wind turbine's cost of the next kWh can be very low, because it doesn't actually need kWh payments to run. Let's say that the wind turbine bids into the real-time market at 0 cents per kWh.

In the real-time market, the rules of dispatch are quite clear: least-expensive plants are dispatched first. So, whatever energy the wind turbine provides will take the place of some of the energy our gas plant was supposed to provide. If the wind energy displaces all the power bid in by the plant, then the price bid in by the next-most-expensive plant will set the real-time clearing price. Let's assume that the result is a new clearing price of 8 cents per kWh.

Hey, but wait a minute! The gas plant has already been paid 10 cents per kWh for that energy, and the wind turbine has not been paid yet. The resolution goes like this: The wind turbine supplies some power; the gas plant supplies whatever power is requested by the grid operator.

The gas plant considers the 10 cents per kWh it received in the day-ahead market and looks at the 8 cents per kWh real-time clearing price. The gas plant returns the money it was paid for the kWh it doesn't have to provide, but it returns it at the rate of the new, lower real-time clearing price of 8 cents per kWh. (The gas plant returns this money to the grid operator, which runs the markets.) In our example, it was paid 10 cents per kWh for energy it doesn't supply but has to return only 8 cents per kWh to the grid operator.

The wind turbine, which bid in at 0, receives the newly lowered real-time clearing price of 8 cents per kWh. When everything has settled, the gas plant keeps the difference between what it received at the day-ahead clearing price and what it pays out at the real-time clearing price.

The gas plant gets a sort of subsidy: it doesn't have to make that energy, but it still manages to make money for promising energy that it doesn't need to supply. The gas plant goes offline, doesn't buy gas, and keeps the difference between the original payment it received and the actual price on the grid. Let's make the example a little more complicated, and say the gas plant does not go offline, but still has to supply some power in the real-time market. The plant keeps the day-ahead payment for the power it does provide, despite the fact that the real-time clearing price on the grid is now lower. That is, it receives 10 cents per kWh for the power it supplied, while the wind turbine gets 8 cents per kWh.

As you can probably guess, it is actually even more complicated than I have described. If the gas plant made an arrangement to purchase just-in-time gas in order to supply energy, and the wind turbine supplied some of the energy the gas plant was planning to supply, the gas plant would have to back out of the gas-purchase arrangement. This may involve paying a penalty. In take-or-pay contracts, common in the gas industry, the buyer has to pay a penalty if it does not take the gas from the supplier.

Still, looking at the overall situation, when wind bids in at the last moment, the gas plant will find itself supplementing its income. The gas plant gets a subsidy (the difference between day-ahead rate and real-time rate) for promising to supply power that it doesn't have to supply, because the power will be made by an intermittent source such as a wind turbine.

In general, due to their operating characteristics, nuclear and coal plants are rarely in a position to supplement their income due to differences in price in the day-ahead and real-time markets.

The market and the partnership

THE GAS INDUSTRY OFTEN sponsors advertisements that show gas as a good partner for renewables. In the ads and in real life, gas turbines back up the output of wind. When the wind's not blowing, who you gonna call? The gas turbine, of course. That is true.

However, those advertisements tell only part of the story. Gas and wind aren't just partners in keeping the lights on; in the RTO areas, they are a mutual-aid society about funding. When the wind energy lowers the real-time energy price, the gas plants make a kind of bonus payment as they shut down for a while.

At one point while writing this book, I made a valiant attempt to describe how the current day-ahead market is bid and later resolved. I looked at the recent Do Not Exceed dispatch rules, which were fully implemented as of June 1, 2019. I acknowledged the fact that a gas plant bids into an electricity market that runs from midnight to midnight, but it buys gas in a gas-supply market that runs from 9 p.m. to 9 p.m. I considered various scenarios in which ISO-NE either overestimates or underestimates the day-ahead demand and how those estimates were resolved. I got exceedingly deep into the weeds.

I finally decided: No. *Shorting the Grid* is not about how to bid into the energy markets if you are a gas plant. Even though the day-ahead market basically favors wind turbines and gas plants, I planned to just skip writing about the whole day-ahead thing. It is too complicated and doesn't add that much to the story of the grid.

Beyond the day-ahead

LOOKING BACK AT THAT decision to skip some writing, I now know that I actually had a simple task at that point. I was trying to describe our current day-ahead market. A one-day-ahead market for kWh. Period.

As it turned out, I could not skip writing about day-ahead, because things happened. Events unfolded. For fuel security, ISO-NE proposed three new markets: a new day-ahead market (ancillary services), a new several-days-ahead market (energy), and a new several-months-ahead market (energy). I had to write about the current day-ahead markets, or it would be impossible to even mention the new proposals.

I can't imagine what the three new proposed markets are going to look like in practice. Three days ahead? Months ahead? Ancillary services? There will be complications. Nobody but the participants will have a chance of understanding the layers upon layers of rules. I can't begin to try to describe this, due to what I learned in my trip into the weeds about the current, relatively simple day-ahead market for kWh.

However, I can definitely imagine what effect these three new markets would have on fuel security in the Northeast.

None.

CHAPTER 22

RELIABILITY AND TRANSMISSION

Not allowed to delist

Remember the headline: "Vermont Yankee is not allowed to delist from the Forward Capacity Auction"? I wrote about this headline in chapter 3. It showed me that I was no longer a maven about the grid. At the time, I had no idea what the headline meant. Now that I know, it's time to explain "not allowed to delist." It's all about reliability.

N minus 1. That phrase describes the main criterion that ISO-NE (and other groups) use for reliability. It means that if the grid is running normally (N), then the failure of any major item on the grid (minus 1) should not keep the rest of the grid from functioning. I used the odd term "item" because the failure could be a major transmission line, a power plant going offline unexpectedly, or a fire at a substation, and so on. In all these cases, the grid is in a minus-1 condition, but it should have enough redundancy to keep functioning as usual. The failure should not spread to become a cascading blackout.

Striving for N-minus-2 reliability is a subject of active research. Doctor Margaret Eppstein of the University of Vermont models 2 and 3 component outages that might lead to a cascading blackout.[88] An even modest-size grid (one of her subjects is the Polish grid) has about 3000 transmission lines. Studying all possible 2-at-a-time failures on the entire system would be an impossible task, even with powerful computers. She uses a random chemistry algorithm (no, I can't explain it) to find which sets of failures will lead to a "malignancy," that is, a cascading blackout.[89]

But back to Vermont Yankee not being allowed to delist.

In 2010, ISO-NE did a reliability analysis for the southern Vermont-southwestern New Hampshire area. ISO-NE concluded that if Vermont Yankee shut down, that area would not meet the N-minus-1 reliability criterion.

However, in August 2010, Vermont Yankee had decided not to bid into the ISO-NE Forward Capacity Auction for the years 2013-2014. If Vermont Yankee had bid into the auction, it would have taken on a "capacity-supply obligation." The good news is that, if the plant was able to run, the plant would have received capacity payments. But, if the plant took on a capacity-supply obligation but was unable to run, ISO could levy a fine. The plant could have conceivably avoided such a fine, if it could find another plant (or several plants) to take on its capacity-supply obligations. However, large power plants are not thick on the ground in New England or anywhere else. And most large plants would already have their own capacity-supply obligations.

Vermont Yankee needed a Certificate of Public Good from the state of Vermont to keep operating after 2012. In early 2010, a small tritium leak had been discovered at the plant This leak caused no danger to the public, but it was a leak. There were protest marches against the plant continuing to operate, politicians made speeches against it, and so forth. Looking at the anger and anti-nuclear crusades of 2010 in Vermont, the plant owner, Entergy, was concerned that the plant would not get its Certificate, and therefore, it would not be able to operate past 2012.

Therefore, Entergy decided not to bid Vermont Yankee into the forward-capacity auction for 2013–2014. The plant might not be operating at that time. The capacity revenues that the plant would receive by bidding into the auction were not high enough to make up for the risk of fines if it wasn't available. (This is my educated guess as to why the owners made this decision. I have no inside information on this subject.)

Vermont Yankee had made the business decision that was best for the plant but not best for the grid. Without Vermont Yankee, a portion of the grid would be below N-minus-1 reliability standards. ISO-NE could not forbid the plant to shut down. But ISO-NE was in a bind, due to the plant dropping out of the Forward Capacity Auction. If the future of the grid shows lack of reliability, ISO is supposed to do something, even with the limited means at its disposal.

We will never know exactly what ISO did at the time, but "not allowing the plant to delist" meant asking the plant to make a business decision that was not in its best interests. There's a way to affect business decisions: money. I think that ISO-NE made a financial deal with the plant so that the plant did not delist. This was an example of an RMR (Reliability Must Run) negotiation. Such negotiations are part of what a grid operator has to do. Of course, other plants hate when a rival plant gets to take part in an RMR negotiation. The rival plant will get more money.

Vermont Yankee closed in 2014, but, by that time, Vermont Yankee going offline no longer led to a reliability problem. In the years between 2010 and 2014, the local power companies had improved their transmission structure, and Vermont Yankee was not needed for reliability.[90] As this history shows, reliability can come through generator availability or through more transmission lines or in many other forms. The RTO does the reliability assessment. In this case, ISO-NE did not have to take any further actions about Vermont Yankee, because the reliability problem had been solved.

The socialized lines of ISO-NE

There are many types of transmission lines. Some are authorized by the grid operator, some shorter ones are owned by transmission companies, and some are built by private, non-grid entities, who plan to make their money by charging for transmission.

Two proposed privately owned lines are central to the controversy about the assessment of ISO-NE's fuel-security study. The Synapse group said that ISO-NE should have assumed that a new line would bring 1000 MW of power from Canada to near Boston. Two lines had been proposed to bring this power.

The Northern Pass line would have come from Hydro Québec in Canada. Then the line would have been routed through New Hampshire, and it would deliver the power to Massachusetts. New Hampshire would not have received any of the power. The line was widely opposed in New Hampshire. In February 2018, New Hampshire's Site Evaluation Committee refused a permit for Northern Pass, and the line probably won't be built.[91]

A rival line, the TDI Consortium line, which would go under Lake Champlain and through Vermont, has a better chance of being built.[92] (There has been a third line proposed, and much of this line would be located in Maine.) These lines would be built by private parties. The RTO does not get heavily involved in siting or paying for these lines.

Since I am writing about the RTOs, however, I will focus on the issues of transmission lines that are required for reliability. These are the lines for which the grid operator does assessments and assigns costs and so forth. The socialized lines.

How do you socialize a line? "Socialized" means that the various distribution utilities pay for transmission in proportion to their share of the system load. For example, let's look at my own little state of Vermont. Vermont is, in general, only about 4% of the power usage in the ISO-NE area. Therefore, Vermont pays only 4% of the cost of transmission-structure upkeep. The costs are "socialized" according to each state's use of the infrastructure. Massachusetts pays much more for the transmission infrastructure than Vermont pays, because Massachusetts has more people and industries.

Note that this payment is completely unrelated to where the lines are located: Vermont may host an important long line, but that doesn't mean that Vermont will be charged more for the upkeep of that line. Similarly, if the grid operator feels that there needs to be a new line in New Hampshire for reliability, New Hampshire will not be charged extra money to pay for that line. Share-of-load pricing will still continue.

Socialization of costs prevents a certain level of NIMBYism about necessary transmission lines. There are plenty of local concerns about lines that bisect farm and forest, but no state can say, "We can't afford that line" about a line needed for reliability. A new line may run through a state, but that state will be paying only its usual share of the socialized costs of the line.

Basically, socialization of transmission line costs has worked because the rules have been clear. In the past, a "socialized" line was built for reliability, and reliability was measured as meeting the N-minus-1 criterion. There's a certain transparency to N minus 1. It gives all the states some trust in the system.

This trust is being eroded. For once, this is not the fault of the grid operators or the power pool. It's a problem caused by FERC, specifically by FERC Order 1000.

Different areas of the country have different rules on socialization of transmission lines. The different sets of rules follow the RTO and non-RTO areas' boundaries, as shown in the map (figure 10) on FERC's Order 1000.[93]

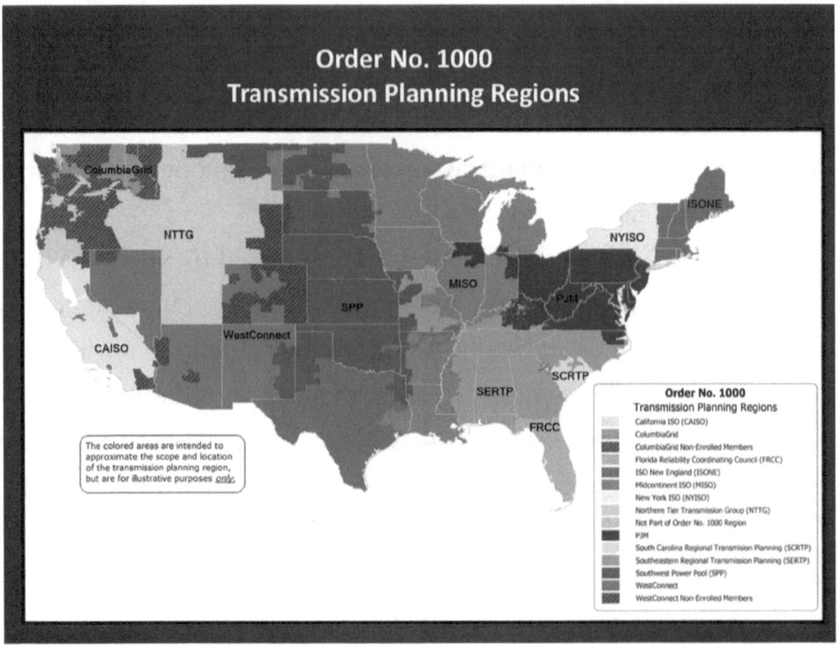

Figure 10: Transmission planning regions (FERC)

The map is a first approximation to the way the rules are set. Actually, the rules are set in response to lawsuits. Congressional Research Service (CRSreport.com) has a lengthy but approachable history of cost allocations (and lawsuits) about transmission.[94]

My book is necessarily limited in scope. I encourage you to investigate how transmission payments are allocated in your own region.

And now, let's look at FERC 1000, which affects the entire country, RTO areas and non-RTO areas alike.

CHAPTER 23

FERC 1000 FLIES UNDER THE RADAR

Transmission: how it gets paid

Because transmission lines link states, the federal government has always been involved in regulating transmission. Paying for transmission has always been a complex issue, from the Otter Tail decision in 1973 until today. In terms of assessing the charge for transmission, however, the background idea is: Those who benefit pay in proportion to the benefit they receive. "Pay for your benefit" is the basis for the socialized costs described in the last chapter.

However, under pressure from the spirit of deregulation, and perhaps more direct pressure from renewable groups, FERC recently changed the rules with Order 1000.[95] Technically, this order went into effect in 2011,[96] but utilities have not rushed to implement it.

In my opinion, FERC Order 1000 will break people's trust in the fairness of transmission-system-cost allocations. FERC 1000 is a huge change, and it has flown under the radar. Most people have never heard of FERC 1000. (FERC 1000 applies to all transmission areas, not just the RTO areas.)

FERC 1000 has two key metrics:

1. Transmission planning must take into account the need for transmission that meets public-policy requirements, not just reliability requirements. Socialized rates must be determined for such transmission-line construction.
2. To encourage competition, an incumbent utility does not have first refusal on competitors' transmission projects in their territories.

Transmission: for public policy

To me, the first one is the most dangerous.

After FERC 1000, the costs of public-policy transmission lines will be socialized between states. If a state wants a transmission line built, not for reliability as assessed by ISO-NE but for public policy, as defined by the state, then other states will pay most of the cost of that line. FERC left the rates to be determined by the various states.

In New England, the state that wants the line for policy reasons will pay 30% of the cost of the transmission line, while other states will share 70% of the costs. (Different RTOs can set their own rules for FERC 1000 socialization.)

If your New England state wants a transmission line to bring power from distant wind turbines to its city center, in the pre-FERC 1000 days, that would be something the ratepayers of that state would pay for. After FERC 1000, one state may decide on a policy, but all states will pay for it.

VELCO is Vermont's electric transmission agency.[97] In a 2013 VELCO meeting, Tom Dunn noted that FERC 1000 will have a negative effect on controlling transmission costs, if the costs due to one state's policies are spread to other states without recourse.[98]

When all states have to pay transmission costs for one state's policies, FERC 1000 will set up a "tragedy of the commons" for transmission. The classic "tragedy of the commons" is when a shared resource is overused and therefore depleted, because each individual user does not have to pay for the effects of his own overuse of the system.[99] The classic example is the use of common grazing land. Unless there is regulation to limit grazing rights, everyone has an incentive to graze many cattle on the common ground, causing the commons to be overgrazed. Future grazing will be degraded.

FERC 1000 sets up a similar problem for transmission costs and a similar level of distrust in the fairness of "socialization." Why shouldn't every state decide to buy some new transmission and meet their renewable goals—at the expense of all the other states? Why should state A pay for state B's goals? And if state A is going to have to pay for state B's goals, State A will try to arrange that State B pays for state A's goals, too.

This is the tragedy of the commons: every state expands its own share of the socialized payments, since all the other states are expanding their share of the payments. More and more money will be required, and the commons will be overgrazed. In this case, "the commons" are the ratepayers. They will find their power bills increasing, and there will be nothing they can do about it.

What about FERC 1000's requirement for competition, though? Isn't that competition going to make things cheaper?

I don't have much time or, frankly, interest in the ins and outs of competitive transmission lines. However, Jason Marshall, of the New England States Committee on Electricity (NESCOE), presented their assessment of FERC Order 1000 and competitive transmission projects. He spoke at a CLG meeting in New England in 2017. He described competitive-bidding scenarios for various types of transmission projects. In many cases, competitive bidding was likely to raise costs. My summary of his talk: "It's complicated. Competitive bidding will not necessarily lower prices for the consumer."[100]

A few words about NESCOE. When FERC set up the RTOs, FERC encouraged states in the RTO regions to form Regional State Committees such as NESCOE. These committees were designed "to coordinate and advance state views in regional electricity markets and planning activities."[101] This was a good suggestion on FERC's part since few individual states could expect to have the resources and the standing to effectively oppose an RTO plan.

Meanwhile, back to competitive transmission. Writing a Request for Proposal (RFP) and evaluating the responses costs money. There are very few bidders who can build a transmission line. Transmission lines are a well-known technology, and the bidders are very likely to bid very similar amounts. For these reasons, many knowledgeable people think competitive bidding will not lower costs but that setting up the competitive RFP will raise costs.

Personally, I think RFPs are an important part of grid governance. RFPs are a partial bulwark (note how I am hedging my words) against crony capitalism. So, I don't totally agree with this argument that RFPs will make transmission lines more expensive. RFPs certainly make things more complicated, though.

Utility Dive's Herman Trabish has a recent article which raises a related topic: Has FERC's landmark transmission-planning effort made transmission building harder? In other words, FERC 1000 may have made transmission planning harder, and not just because of RFPs.[102]

Did FERC 1000 have this effect? Probably. Most of the Trabish article is about lawsuits and who has jurisdiction. Still, in that article, former FERC Commissioner Tony Clark is quoted as saying that

Order 1000 imposes too much of a regulatory burden. Clark also notes that two sections of the Order mandate regional planning, while another section mandates competitive bidding for transmission lines. Competitive bidding and regional planning are not easy to meld together, though it is possible.

In Clark's opinion, Order 1000 goes off in too many directions at once, and it has impeded the building of transmission lines. I agree with Clark. As a matter of fact, I agree with Confucius, who said that a man who chases two rabbits will catch neither.

The effect of FERC 1000

FERC 1000 HAS ALREADY caused the filing of many lawsuits. And it will no doubt lead to much regional rule making, which will also lead to more lawsuits. It is not clear what consumers will gain from getting this order in place.

According to the NESCOE presentation by Jason Marshall,[103] former commissioner Clark wrote that:

> Panelists described Order 1000 as "either partially a success, generally a failure, cost containment is either important, unnecessary, or unneeded, should be flexible or very well-defined."[104]

Ultimately, ratepayers are paying the lawyers for all of this.

I hate it when the ratepayers get to pay, but we don't get much say.

At the time of Marshall's presentation (2017), I was a member of the Consumer Liaison Group Coordinating Committee at ISO-NE. As a member of that committee, I suggested FERC 1000 should be a topic for one of our public meetings, and it was. It hasn't been the topic of many public meetings, though it is a revolution (or could be) in transmission charges. The FERC 1000 order has flown under the radar. In my opinion, this means that the ratepayers will suffer.

In my opinion, "under the radar" and "suffering" can be said of many rules on the grid. Still, FERC 1000 has more sweeping implications than most rules. It needs to be carefully defined or eliminated. I think "eliminated" would be better.

CHAPTER 24

THE NOT-STRESSED GRID IN SUMMER

The hottest day and its greenwashing

IN CHAPTER 22, Reliability and Transmission, I described how the costs of socialized lines were assigned to states. In that way, even if a transmission line was needed for reliability, and it had to be built in Vermont, Vermont would only pay about 4% of its costs or its upkeep, because Vermont is only 4% of the New England load.

Actually, ISO-NE doesn't assign transmission costs to states. It assigns such costs to distribution utilities, according to their share of the load. For example, in Vermont, the socialized transmission costs are shared between Green Mountain Power, Vermont Electric Cooperative, and several smaller co-ops and municipal utilities.

Transmission costs are rising as a portion of the customer's bill, as figure 11 shows.

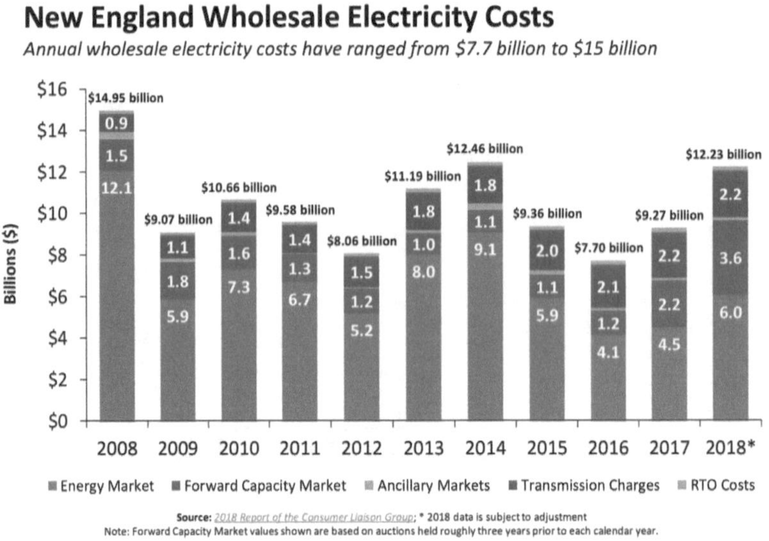

Figure 11: New England wholesale electricity costs, showing rise of transmission costs. (ISONE)

The chart above is from the June 2019 presentation of ISO-NE to the Consumer Liaison Group,[105] and is based on data from the 2018 annual report.[106]

As you can see, transmission costs have increased significantly in the last ten years. Transmission was about $1 billion in 2008 and part of a total wholesale electricity cost of $15 billion. In 2017, the last year for which we have complete data, transmission was more than $2 billion, though total wholesale electricity costs had fallen deeply, to $9 billion. An alternative way to view the changes is that transmission costs changed from 6% of total cost in 2008 to 24% in 2017.

Utilities in RTO areas have a new, systematic way of taking advantage of consumers' good intentions. Utilities misrepresent their transmission-cost strategies.

We all want to be clean and green. We are all aware that hot days mean lots of air conditioning and lots of power usage. It stands to reason that hot days stress the grid (it may be reasonable, but it's not always true). We all want to do our civic duty and try to relieve that stress. Utilities take advantage of this and recruit us to help with cost-shifting.

In the Northeast, utilities publicize their "beat the peak" days. Some utilities ask their consumers to cut back on their air-conditioning and appliance use when they expect peak power usage on the grid. This request often comes with the simultaneous and somewhat contradictory advice to keep safe and cool, and especially make sure that old people, children, sick people, and pets don't get overheated. Meanwhile, other utilities crow that their battery installations have "beat the peak" and allowed them to save, say, $600,000 for their customers.[107]

Very few utilities admit that this effort is not about conservation. If you read the complete Green Mountain Power press release, all the way to the bottom, you see that Green Mountain Power admits that ISO-NE looks at the peak hour to calculate the costs that utilities pay for the regional grid at the peak hour. Utilities can save money for their customers by saving energy during the peak hour. This isn't about conservation, but that is hidden under "green" rhetoric. Nowhere in the description does Green Mountain Power admit that, if it succeeds at "saving" $600,000, then other utilities are going to pay that $600,000. The cost of maintaining the grid does not change because some batteries were used during peak hours. Only the allocation of the cost changes.

The verbiage above the "$600,000 savings" statement in the press release is about saving carbon (during the peak hour), the equivalent of not burning gasoline, and so on. It is written as if "beating the peak" is about saving energy and being green. Actually, no. The batteries are not particularly effective in saving energy costs, but they are tremendously effective about shifting transmission costs to other utilities.

Let's look a little more closely at the Northeast grid in summer.[108]

Electricity in the Northeast

IN EARLIER CHAPTERS, we looked at winter on the Northeastern grid. Oil saved the grid—but it almost failed at saving the grid. Rolling blackouts are likely during winters to come.

The Northeastern grid is heavily dependent on natural gas, which gets delivered by pipeline. When homes and businesses are using natural gas for heating, power plants sometimes cannot get gas. The power plants often use oil instead, but sometimes the winter weather is so bad that it is difficult to deliver the oil. I have several chapters on the problems of just-in-time gas delivery and questionable oil delivery. Winter is tough in New England.

Contrast this with summer. In summer, natural gas is readily available, and the New England grid is not particularly stressed.

Note: I must mention that the grid can be stressed at any time when too many plants go offline, for whatever reason. On Labor Day, 2018, a major gas-fired plant went offline, and prices spiked to $2400/MWh in New England.[109] Higher-than-expected temperatures plus a major power-plant outage led to this situation. ISO-NE had to buy power from neighbors, but they did not have to ask consumers to cut back on their power use. Basically, because of natural-gas availability, the crisis was of short duration.[110]

Early in the summer of 2018, I took note of a major heat wave. How could I not take note of it? Like most people who live in Vermont, we don't have central air conditioning. We have fans, and we have two underpowered window air conditioners in upstairs rooms. Yes, I noticed the heat. However, this book is about the grid, and I also noticed the grid. The grid was doing fine. (My blog post,[111] written at the time, shows fuel use and demand during that heat wave.)

How can I say the local grid was not stressed in hot weather? We were having a major heat wave. For days, Vermont temperatures had been in the high nineties. A number of communities opened "Cooling

Stations" in public buildings such as fire departments. People were encouraged to go to air-conditioned malls, drink water, and to check in on elderly people who may need assistance.[112]

Okay, it was very hot. But the grid was not particularly stressed.

I will start by comparing the grid situation in this heat wave with the grid situation in the cold snap in December–January.

A typical hot day on the grid was July 3, 2018. The peak was near 25,000 MW, which was pretty high. However, the prices were not high. The LMP (local marginal pricing) wholesale prices for electricity were between about $25 and $80 per MWh, or about 3 cents to 8 cents per kWh.

What about fuel use? Using more electricity in New England always means making more carbon dioxide and burning more fossil fuels. So, conservation is always good.

But is conservation in summer particularly wonderful? Not really.

On a typical hot day, July 3, 2018, we had a fairly clean grid. The fuel mix was mostly natural gas, nuclear, hydro, and renewables (solar, wind, biomass). The grid was running at 60% gas, 20% nuclear, 16% hydro and renewables. Pretty good, in terms of emissions.

In contrast, during the cold snap at the beginning of 2018, electricity use never got much higher than 22,000 MW, but LMP prices spent a lot of time between $150 and $300 per MWh (15 cents to 30 cents per kWh). In the winter, when natural gas was not available, oil and coal were in heavy use. During the cold snap, the mix was 30% oil, not "less than 1%" oil, as it is now. Coal use was higher, also, up around 5%. Though we were using less electricity in the winter, we had a more carbon-heavy mix of fuels.

In terms of emissions, there's no particular reason to conserve in the summer instead of conserving at some other time. But let's look at something else. Even with natural gas available, wouldn't the grid be close to maximum capacity in hot weather? If we don't conserve, wouldn't the grid fail?

Nope. The grid was doing well. The ISO-NE website lists "surplus capacity" right on the front page. As I was writing my blog post on a July 3, surplus capacity on the grid was 1,180 MW.

What is surplus capacity? It is the capacity available above the maximum predicted peak power use for day (23,000 MW), and it is also above the grid's operating reserve requirement for the day (2,492 MW). In other words, the grid was expected to use 23,000 MW, and capacity in reserve was 3,672 MW. There was a 16% margin above expected peak demand. The grid was looking fine and stable.

I did a careful analysis to reach the conclusions above. I could have skipped doing my own analysis. I could have simply noted that, when ISO-NE studied the reliability of the grid and modeled many scenarios showing rolling blackouts in the future, ISO-NE looked at *winter* stresses on the grid. (See chapters 19 and 20 of this book.) ISO-NE was not concerned with high summer electricity usage leading to blackouts. Their examples were from the stressed days of winter.

In short, conserving electricity in a New England summer doesn't save more money or more carbon dioxide than it would save at many other times. As a matter of fact, it saves less of both than it would save in a winter cold snap, when oil usage is high. Conserving in summer also doesn't "save the grid" from blackouts. The grid is operating at high capacity but has plenty of power in reserve.

So, why are the New England utilities pushing conservation in the summer?

The utilities are urging conservation in summer because they are playing the Game of Peaks. It's a utility game about money. If they play the Game of Peaks well, they can shift some costs from themselves over to neighboring utilities. Yeah, it's a zero-sum game ("I win" can happen only if "you lose"). Let's look at the rules for that game.

CHAPTER 25

THE GAME OF PEAKS

CUTTING BACK ON ELECTRICITY use on the hottest day of the summer is not a moral imperative. It is merely part of The Game of Peaks. This game allows large utilities to shift costs to smaller utilities and co-operatives.

Luckily Game of Peaks is all about accountants, not swords. It's nowhere near as brutal as the Game of Thrones. Nobody gets killed in the Game of Peaks, but lots of people get misled about the situation on the grid. And lots of people end up paying more than their fair share of grid costs. There are losers in the Game of Peaks. You may be one of them.

ISO-NE sets the rules for the Game of Peaks

ISO-NE MUST CHARGE utilities their "fair share" of system costs, particularly transmission costs. But what is their "fair share"? ISO-NE determines a utility's share of the grid-wide transmission costs by determining the power used by that utility during the peak-usage hour on the grid. (It's a little more complicated, of course, but that is the basic method that ISO-NE uses to determine how much different utilities pay for maintaining the transmission system.) Basically, the percentage of power a utility uses during the peak hour is the percentage of transmission costs that the utility has to pay.

If a utility can lower its electricity use for that one peak hour, it will save a lot of money by paying lower transmission costs for the grid.

This percentage-of-load calculation is an opportunity for utilities to shift costs elsewhere. Utilities campaign about "shaving the peak." Announcements state that "we saved hundreds of thousands of dollars by shaving the peak."

For example, in a *Burlington Free Press* article from 2016, Green Mountain Power claims to have used batteries to reduce its peak power demand, saving customers $200,000 in an hour.[113] In early 2018, Green Mountain Power claimed to save its customers $500,000 by peak shaving during a July heatwave.[114]

Though Green Mountain Power "beat the peak" in July 2018, it turned out that the July day was not the peak. So much for that press release. August had even hotter and more humid weather. In August, Green Mountain Power announced that they had saved $600,000 for customers by beating the new version of the yearly peak.[115]

Conservation now?

THESE STATEMENTS ABOUT saving $200,000–$600,000 in an hour are misleading. The statements are phrased as if energy conservation at the peak is about sparing the grid, avoiding the use of fossil fuels, and so on.... That isn't what "beat the peak" is about.

When you read a "Beat the Peak" article, the first part of the article is usually about preventing the use of fossil fuels. For example, in Green Mountain Power's August press release, the first part is about saving money and carbon with battery storage. Next, Green Mountain Power writes "the amount of stored energy deployed to beat the new peak last week was like taking up to 6,000 homes off the grid, and it offset about 21,120 pounds of carbon, the equivalent of not burning 1,078 gallons of gasoline."

In fairness to Green Mountain Power, near the bottom of the press release, they do describe the event in a more complete manner. "ISO-NE uses the yearly peak hour to calculate costs utilities pay toward the regional grid, so when utilities are able to lower demand during that key hour, they can create savings for customers." Not all utilities are that honest.

The multi-hundred-thousand-dollar savings ($200k to $600k, depending on the year) is not the excess cost of electricity in that single hour. The savings are due to the fact that Green Mountain Power used its predictive power and its batteries to reduce its demand at the time of peak demand. Therefore, it will reduce the amount it pays for grid-level transmission.

Somebody is still paying that $200k for transmission: the overall cost of grid transmission hasn't changed. Some other utility is paying that cost. As Green Mountain Power describes it: "... when utilities can lower demand during that key hour, they can create savings for customers."

In other words, "beating the peak" provides savings for Green Mountain Power customers, not savings for the grid.

Saving electricity in summer: the game as played

ACCORDING TO ITS July 2018 press release, Green Mountain Power now has 5,000 kWh of battery storage.[116] Their 5 MWh of storage will not make much difference to the expense of transmission on the grid. As discussed in the last chapter (the Not-Stressed Grid in Summer), the New England grid runs about 10,000 MW at night and up to around 25,000 MW during a hot day. 5 MWh storage is pretty small.

However, Green Mountain Power expected that storage would make a major difference to their own bottom line, and it did.

"Beating the peak" is not about:

- saving money while the grid power is expensive (it is not that expensive in summer) or
- diminishing pollution more than usual (coal and oil are not in use much during the summer) or
- keeping the grid from failing (there's plenty of reserve capacity in summer on most grids, including New England)

Utilities advertise about the grid when they want to "beat the peak." By careful description (cars off the road, etc.) they make "beating the peak" sound like a moral imperative. It isn't.

Unfortunately, people know very little about the grid, except that you "shouldn't" use as much electricity on a hot day in summer.

For most grids, that belief is based on a fallacy. If I am going to write about the difficulties an RTO faces in keeping the lights on in winter, I think I need to also write about the misleading rhetoric of utilities in summer. (The Texas grid is an exception. It has low reserves and can have problems in the summer.)

Don't get me wrong. Being thrifty and not using excess power is always a good thing. Still, especially if you live in New England, it helps the environment more if you are thrifty with electric usage in winter (with all that oil and coal-burning) than in midsummer. It helps your local utility's bottom line more if you are thrifty with electric use in summer.

The number of people who will read this book is probably small compared to the number who will hear a utility public-service announcement about Beating the Peak. Still, I wanted to clear this up.

Utilities are masters of greenwashing: The official announcement encourages you to help the environment. If you could hear the little man behind the curtain, he would be saying, "Do this to help my company's bottom line."

RTOs and transmission

As far as I can tell, RTOs do as good a job with transmission as non-RTO areas. After all, transmission concerns led to the formation of RTOs.

There are exceptions, however. PG&E in California is part of the California RTO system, and PG&E has filed for bankruptcy due to its extensive liability for wildfires caused by allegedly poor maintenance of its transmission lines. According to a January 2019 article in *Fortune Magazine*, PG&E's wildfire liabilities may top $30 billion.[117]

I can't even begin to answer the question of whether PG&E's line-maintenance failings had anything to do with the fact that the company is in an RTO area.

As a matter of fact, PG&E operates more like a vertically integrated utility (it owns both power plants and transmission lines) than a distribution utility or merchant generator in an RTO area. In RTO areas, utilities are supposed to be either regulated "distribution utilities" who own power lines, or merchant generators, who supply kWh to the distribution utilities. In reality, the two roles are often mixed, as they were with PG&E.

The jury is still out on how this disaster happened. And clearly, there will be more than one jury involved. I do not think my speculations on the California situation will lead to any particular understanding of transmission lines in the RTO areas, and so I will not speculate. I did need to mention the California situation, murky as it is. The resolution of the California claims will influence transmission requirements everywhere. Nobody can say how they will be influenced, but they will be influenced.

RENEWABLES ON THE GRID

CHAPTER 26

RENEWABLES, REALITY, AND ELECTRONS

The state of the states

TO TALK ABOUT RENEWABLES, we have to look at state policies about renewables. RTOs are supposed to be fuel neutral, but state policies have no such constraints.

Federal policies can affect renewables. As noted in a previous section, the federal government, through FERC 1000, allows states to force their neighboring states to pay for the state-level policy choices. This is taxation without representation: the people in the neighboring states don't get to elect the people who set the policy, but the people in the neighboring states have to pay for the policy. So far, we have looked at RTO policies and FERC policies. What about state policies? Are their policies reasonable? Are some of their policies even possible?

There are many ways that state policies can affect the grid: for example, a state may put a moratorium on new gas pipelines or coal plants or nuclear plants in the state. In recent years, state renewable mandates have had the biggest effect on the grid. In this chapter, we will look at the effect of renewables on electricity generation and how the grid operator must work around issues caused by intermittent renewables. In later chapters, we will look at payments and policies about renewables. But for now, it's all about the electrons.

The main purpose of every grid is to make it possible for people to flip a switch and turn the lights on. How do renewables affect the actual operation of the grid?

What is baseload?

MANY RENEWABLE ADVOCATES like to say that baseload is an outmoded concept and that we don't need baseload plants anymore. This is cleverly stated but not accurate.

Baseload is the electricity demand that is in place 24 hours a day, seven days a week. For example, in April 2016, I took a screenshot of the demand on our local New England grid. It was a mild spring day. The steady demand (24-7) was 10,000 MW. During the day, the demand rose to about 15,000 MW. (The screenshot is figure 1 in chapter 4, "The Angelic Miracle of the Grid.")

In the periods of high demand, these numbers can be higher. For example, it was below zero on the night of January 14, 2019. Many people run electric space heaters when the temperature dips. ISO-NE shows that the lowest demand on the grid that night was around 12,000 MW, rising to about 18,000 MW during the day.

Similarly, Labor Day 2018 was hotter and more humid than it was expected to be. When a major gas-fired plant went offline, the grid prices soared. ISO-NE wrote a report about the incident.[118] The minimum load that day, at about 4 a.m, was 12,000 MW. The maximum load, near 7 p.m. was 23,000 MW.

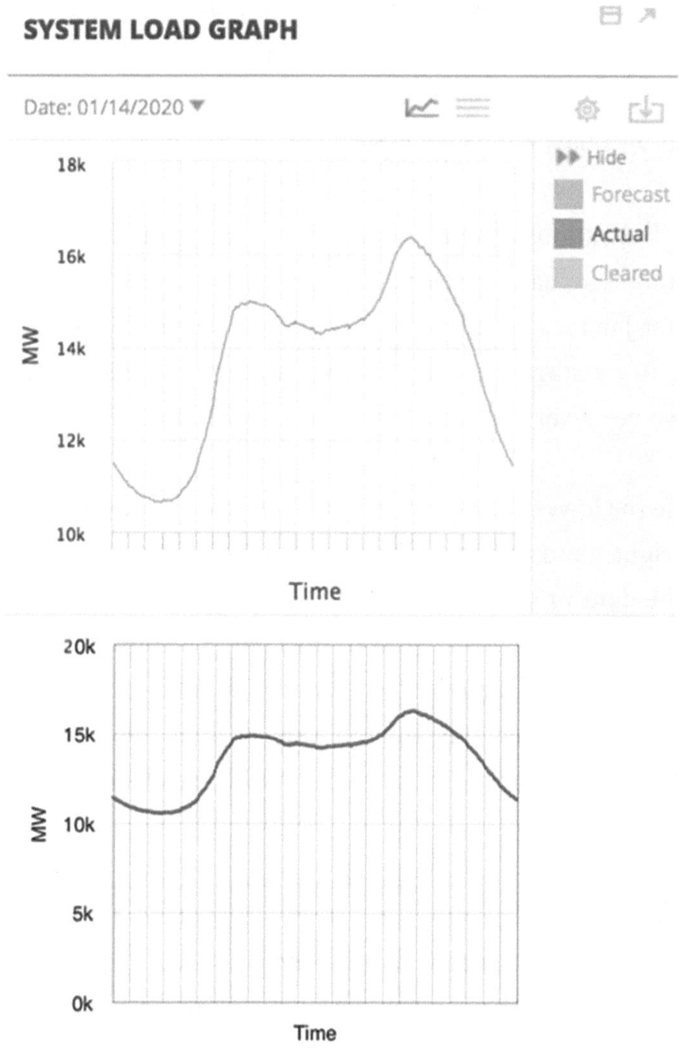

Figure 12: Graphs of actual demand on the ISO-NE system, January 14, 2019. (ISO-NE and George Angwin)

In other words, 10,000 MW of power needed at night on a mild spring day is a true baseload. The grid has to supply that power, all the time. This is not an outmoded concept.

However, for an illustration of baseload, let's look at the colder day, January 14.[119] The April day had power demand that varied between

10kMW and 15k MW. Using such a mild spring day for the graphic could look like a "cleverly chosen example that shows little variation in demand." The January 14 electricity usage varied between 12k and 18k MW, and will be a better example.

ISO-NE's illustrations of demand are misleading about the amount of electricity that is actually baseload. The top part of figure 12 is from ISO-Express for January 14, 2019. The variation in demand looks huge.

However, if we start the demand axis at zero (no demand for electricity), we get a very different view, as shown in the lower part of the figure.[120]

To generate the lower curve, I started by using the download arrow in the upper-right-hand portion of the upper image. This generated a CSV file of the data in the upper part of the image. In other words, the ISO-NE chart and the lower part of my chart both use the same data. In both charts, the time axis starts at midnight. In the lower part of the chart, however, the demand axis starts at zero, and the chart explicitly shows that most of the demand is baseload, even on a high-usage day.

Baseload plants

Traditionally, baseload power has been provided by "baseload plants:" These are plants that are very good at a steady, reliable, inexpensive operation. In general, baseload plants are steam plants: nuclear (my favorite) and coal (there are a lot more coal plants than nuclear in the world). The electricity demand that ramps up during the day and lowers late in the evening is generally provided by "load-following plants." These are more expensive to run but more flexible in following the load. They tend to be gas-fired plants and hydro. However, in France, nuclear plants are also used for load following.

When designing a plant, the designer optimizes for flexibility or for steady output. Of course, a steady-output plant can be somewhat flexible, and a flexible plant can be run in steady mode, but they are not optimized for these situations.

Instead of thinking about power plants, let's think about diesel generators. If I were going to design a stationary diesel generator, for example, as a back-up generator at a hospital, I might not care very much about how much the machine weighs. I probably won't spend much money to make it lighter, as long as it is light enough to deliver and install. On the other hand, if I were going to build a prototype of a new diesel truck, I would care a great deal about how much the engine weighs. A weight-efficient engine will increase the truck's payload and make it a better truck. I will be willing to spend some money to reduce the weight of the engine, if I can do so without hurting the truck's reliability.

Different uses mean different engineering constraints. The materials of my cozy home would not be appropriate materials for a pier on a stormy coast.

As you can see, we need baseload power: on the Northeastern grid, we need about 10,000 MW of power available all the time. Energy efficiency might lower that number; growing populations or regularly charging large numbers of electric vehicles at night might raise the number, but there is going to be some level of baseload demand.

What renewable advocates mean by "We don't need baseload" isn't that they believe all electricity usage will cease at night. I live in Vermont, and I would really hate it if my furnace controller or heat pump (if I had one) wouldn't work at night, when winter temperatures often dip below zero.

What the renewable advocates mean by "We don't need baseload" is that we no longer have to go through the engineering discipline of designing some plants for baseload use and others for load-following use. It is sort of like saying, "We don't need big trucks to carry goods down the road. They have very poor acceleration. They are not flexible. All we really need is sports cars, for everything."

The general idea here is that renewables will do it all. They won't, but people want to believe that they will. Pro-renewables advocates will imagine complex scenarios with inexpensive grid-level battery storage (which doesn't exist yet and may never exist) or thermal storage and then making electricity from the hot fluid that you have carefully stored in some insulated cavern, and so on. "It could work."

Looking out for "could"

YOU HAVE TO WATCH OUT for that word "could." They "could" build a bridge across the Connecticut River every two miles, and I wouldn't have to drive so far to cross from Vermont into New Hampshire. But nobody is going to do that, because it would be very wasteful. But they "could."

Will renewables do it all? Not really. Renewables are designed to work however they work. They are not optimized for a certain service on the grid. For example, ridgeline wind around here is more available at night. However, "more likely" to be available at night does not mean "you can count on wind power at night." When people discuss how renewables "could" do it all, they generally describe a whole string of things that "could" happen, from advanced batteries to a new, huge set of continent-spanning high-voltage DC lines. "Could."

What actually happens isn't "could." What actually happens is that we build the most efficient and cost-effective wind turbine we can build, and then it operates when the wind is blowing. Sometimes wind makes a lot of energy at night, and sometimes it doesn't. Wind doesn't make baseload power, and it doesn't make load-following power. It blows when it is blowing, and the rest of the grid has to work around it. Wind is truly the poster child for "intermittent power."

How will we use renewables? Most of the time, the grid has to work around them.

CHAPTER 27

WORKING AROUND RENEWABLES

Different kinds of renewables

NOT ALL RENEWABLES ARE intermittent. Some types of power are classified as renewable, and they are basically heat engines, very similar to other heat engines that use coal or gas or uranium. In my area, biomass (mostly wood-burning plants) and waste-to-heat plants are both classified as "renewable." These are most often steam plants using fuel that is classified as "renewable." (Some waste-to-heat plants gasify their fuel and then use internal-combustion engines.) Looking at the fuel classification, it is clear that trees grow, and so biomass is renewable if you don't over-harvest the forest. I have always had an issue with waste-to-heat as renewable. Is garbage a renewable resource for energy? Aren't we supposed to be recycling more and making less waste?

I have no real objection to waste-to-heat, though I am not sure the renewable classification truly fits this source of energy. However, I find I can easily surprise people with simple diagrams about renewables from ISO-NE. Around here, biomass is often about 40% of our renewable energy on the grid, and waste-to-heat is another 40%. When the wind is blowing strongly, wind can be 40% or even 50% of renewable energy. In that case, the percentage of renewables on the grid increases with the extra wind energy, and the percentage of biomass and waste heat shrink as a percentage of renewables. Overall, however, biomass and waste-to-heat are the steady performers.

How many times have you seen a report extolling the renewable future, and it features an illustration of a waste-to-energy plant? Never, I suspect. It's always a solar panel or a wind turbine on the cover.

Figure 13 shows fuel use on a mild July afternoon at ISO-NE (July 8, 2019). There were 6% renewables on the grid (upper chart), 64% of the renewables were biomass and refuse (lower chart), and 31% were wind and solar.[121] You can see that what you usually think of as renewables provided 2% of the power in New England. Renewables were 6% of the energy, and the poster children (solar plus wind) were one-third of that energy.

In chapter 9, "Oil Saves the Grid," about a winter-distressed grid in January 2018, you can see a similar chart (figure 4). Renewables were 6 to 9% of the generation on the grid. According to the ISO-NE Cold Weather Operations report at the time,[122] wind performed well most of the time, but solar output was reduced by the fact that most of New England received eight to eighteen inches of snow in the storm. Though not mentioned in the ISO-NE report, the refuse and biomass plants basically continued to chug along as usual.

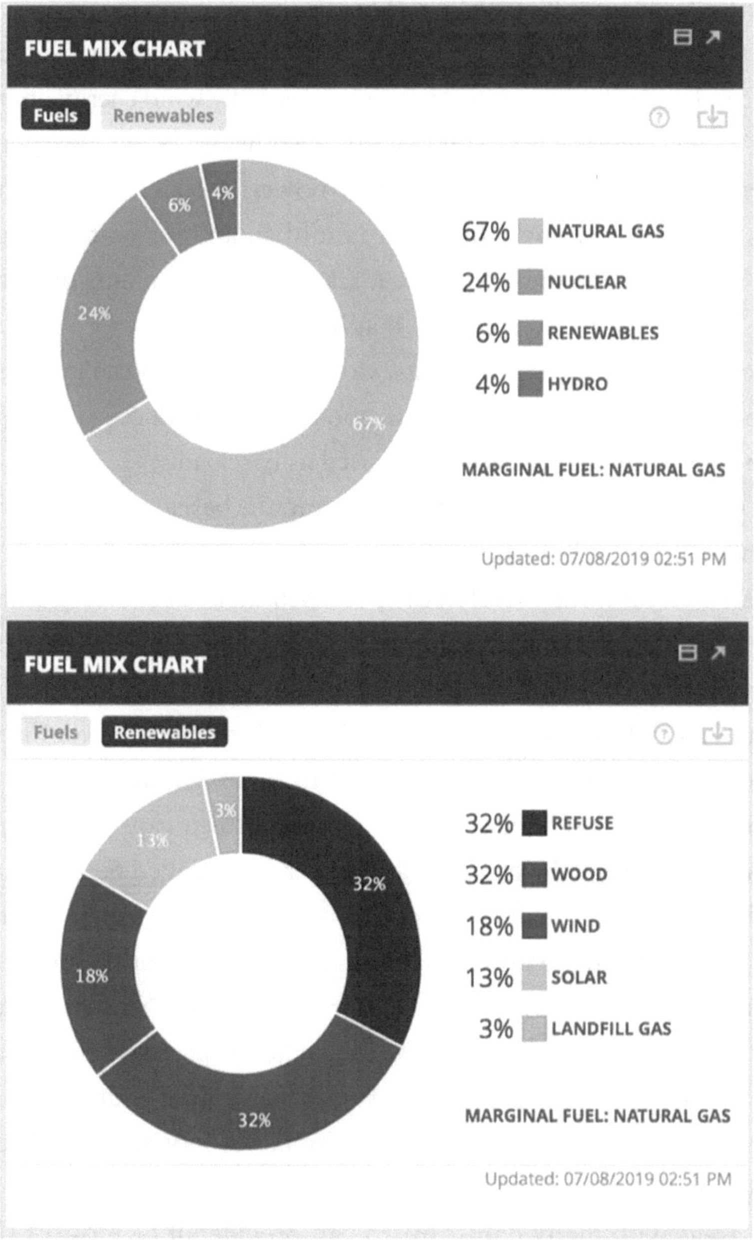

Figure 13: Fuel mix charts for all fuels and for renewables only, updated 2:51 pm, July 8, 2019 (ISO-NE)

Renewables: intermittent or steady

WIND TURBINES AND solar panels are the intermittent renewables. They can be significant sources of energy, especially wind energy on the Great Plains. Here is a brief summary of the other types of renewables.

Hydro power, biomass plants, cow power (methane from dairy farms), landfill gas (methane from landfills), and waste-to-energy plants all operate on the grid in the manner that other plants operate. Some are baseload, and some are load-following.

Biomass plants are steam plants. Cow power and landfill gas must use the methane gas more or less as soon as it is produced. Biomass, cow power, and landfill gas are best used as baseload.

In general, a standard hydro plant cannot be baseload because the turbines cannot run all the time. To operate, the hydro power plant requires a certain water level in the pond behind it. The level of water in the pond behind the dam sinks as water goes through the turbines. When the turbines stop running, the level of water rises as the streams refill the pond. Hydro is generally used for load following, or for filling in when other renewables (such as wind turbines) go offline.

Power plants of all kinds can be described by their capacity factors: the electrical energy the plant produces over an extended period compared to the electrical energy it could have produced at full operation during that period. No power plant has a capacity factor of 100%; at minimum, they all need some outages for maintenance. Nuclear plants' capacity factors average above 93%, the highest capacity factor of any type of power plant. The average capacity factor for a hydro plant in America is about 40%, and this capacity factor mostly depends on the availability of water.[123]

In areas with many dams, hydro can provide all or most of the electricity required, including base load. While a single hydro plant is limited by its capacity factor, using several hydro plants sequentially can provide baseload power.

There are also run-of-the-river dams that do not have significant ponds and are designed to make power whenever the river is flowing. Many large run-of-the-river dams take advantage of a separate upstream dam to provide an even flow through their hydro plant. For example, the Chief Joseph Dam on the Columbia River is a run-of-the-river dam, with only a small pond. In order to operate steadily, it depends on a steady supply of water from the upstream Grand Coulee Dam.[124] This type of run-of-the-river hydro is basically another way to use several dams sequentially to provide baseload.

Hard-core renewables

WIND AND SOLAR ARE the technologies that most people think about when they think of "renewables." Indeed, many hard-core renewables advocates accept only solar, wind, and (sometimes) hydro as renewables. Biomass rarely makes the cut as a true renewable. Professor Mark Z. Jacobson of Stanford plans WWS (Wind Water Solar) as the energy sources for the world. In 2015, Jacobson and others published an article in the *Proceedings of the National Academy of Sciences* on using WWS for all purposes.[125]

In 2017, a group of professors headed by Christopher Clack responded with an evaluation article also in the *Proceedings*.[126] The first paragraph of the Clack article stated that "We find that their (Jacobson analysis) involves errors, inappropriate methods, and implausible assumptions." For example, their rebuttal paper pointed out that the Jacobson paper describes hydro power as providing 700 to 1300 GW. However, existing installed hydro capacity is 87 or 145 GW, depending on whether pumped hydro is included, and the most useful sites have already been exploited.[127]

When the Clack paper appeared, Jacobson published a letter in the same issue of the *Proceedings*, claiming "The premise and all error claims (of the Clack paper) … are demonstrably false."[128] Jacobson said that his assertion on the availability of hydro power was an "assumption," not an error. As Jacobson wrote in the published letter: "The value of 1,300 GW is correct, because turbines were assumed added to existing reservoirs to increase their peak instantaneous discharge rate without increasing their average annual energy consumption." Shortly after the Clack paper and the Jacobson rebuttal were published in the *Proceedings*, Jacobson sued Clack and the *Proceedings* for defamation.

Jacobson later dropped his lawsuit. On the Greentech Media website, Julian Spector wrote an article about the controversy and the lawsuit.[129] In his article, Spector notes that "this 'assumption' (about hydro) was unwritten" in the original Jacobson article. In other words, in his original paper, Jacobson did not describe his assumption that multiple turbines would be added to existing dams. Frankly, adding about ten times as many turbines to existing powerhouses seems very unlikely to me. Dam construction is a massive undertaking. Putting many more turbines in an existing powerhouse ... well, I can't see how that could even work.[130]

Jacobson did drop his lawsuit, which should be a happy ending, I suppose. However, many people, including myself, feel that the fact that Jacobson even brought a lawsuit has had a chilling effect on the whole renewable-energy debate. If scientists can't debate each other in peer-reviewed journals without fear of lawsuits, science will not be able to move forward very well.

There are two books directly refuting the Jacobson plan. *Roadmap to Nowhere: The Myth of Powering the Nation With Renewable Energy* by Mike Conley and Tim Maloney is available as a free PDF download on the web.[131] Mathijs Beckers, of the Netherlands, wrote *The Non-Solutions Project,* available as an ebook or paperback.[132] The work of these authors is clear and easy to follow.

Jacobson writes about using only Wind, Water, and Solar. Other renewable advocates don't even want water (large hydro) included in the mix. In January 2019, more than 600 organizations signed a letter urging "legislation to address the urgent threat of climate change."[133] These organizations include Greenpeace USA, 350.org groups, Eco-Poetry.org. and the Central New Jersey Coalition Against Endless War (which ceased holding vigils in 2015). The open letter is related to the Green New Deal movement, but I don't mean to imply that the letter expresses that movement's exact policy. The Green New Deal is an evolving concept, and the open letter expresses one version of it:

> As the United States shifts away from fossil fuels, we must simultaneously ramp up energy efficiency and transition to clean, renewable energy ... in addition to excluding fossil fuels, any definition of renewable energy must also exclude all combustion-based power generation, nuclear, biomass energy, large-scale hydro and waste-to-energy technologies.

In other words, reliable renewable power sources would not be allowed in this proposed renewable future. This open letter does not propose an effective way to plan a grid, no matter how many organizations have signed it.

CHAPTER 28

THREE ISSUES WITH RENEWABLES

THERE ARE THREE MAJOR problems with integrating intermittent renewables on the grid. One is their "spikiness," a second is their reliability, and the third is their effect on the power supply itself.

Daily spikiness needs backup

BY "SPIKINESS," I MEAN that clouds pass across the sun, and the wind starts and stops when it wants to. The output from renewable energy can go up and down quite quickly. To keep the grid in balance, something else has to be ready to start up quickly when a renewable gets spiky. In general, the thing that starts up quickly is a gas-fired plant that is deliberately kept in a state where it is ready to run. In 2016, the National Bureau of Economic Research (NBER) showed a tight relationship between natural-gas plants and renewables.[134] The *Washington Post* reported on the study:[135]

> All other things equal, a 1% percent increase in the share of fast-reacting fossil technologies is associated with a 0.88% percent increase in renewable generation capacity in the long term.

These results come from observations in more than 26 separate countries, over more than two decades.

As described in the *Washington Post* article about the study, most energy experts agree with these results. Some experts are not too worried about these gas plants, since, after all, they don't run all the time, so they will not add "that much" carbon dioxide to the atmosphere. On the other hand, a wind turbine on a ridge in New England has a capacity factor of about 30%, and the wind blows more strongly at night. So, the gas plant could run as much as 70% of the time.

Whether or not they are concerned with the carbon content, the experts interviewed by the *Washington Post* all agree that the cost of renewables should include some allowance for the cost of the backup plants.

Meanwhile, how will a grid operator deal with this spikiness? Basically, the operator will arrange for fast-startup plants to be available for backup. However, simply having such a plant on the grid does not necessarily provide fast backup for the renewable's spikiness. The plant must also be ready to begin operation very quickly. This often means keeping the plant running at a low level, or keeping the turbine spinning without a load, or various other ways to be sure that the plant can come up to speed quickly. Keeping the backup plant ready to start quickly may require that the plant burn some fuel. The cost of such fuel will be part of "ancillary services" paid for by the grid operator, since the backup plant's readiness is part of grid reliability.

Lack of daily reliability, need for backup power

THERE ARE STUDIES THAT show renewables could provide all our power. Such studies assume grid-level battery storage (that doesn't exist), or they assume that fifteen times the amount of hydro will be available from existing plants (not possible, even if some studies make this assumption), or they assume a huge buildout of cross-continent, direct-current, high-voltage transmission lines. Instead of even attempting to analyze the various kinds of "we could" studies, let's look at how real renewables affect the real grid.

1. The need for backup power: worldwide picture—the NBER study

The NBER worldwide macro-study described above looked at how renewables affect the grid. This study showed that the grid needs slightly more fast-reacting fossil available than it has intermittent renewables installed. (1.0 MW fossil installation is needed for every 0.88 renewable installation.) This macro-study looked at many countries over two decades.

2. The need for backup power: small utility, Vermont Electric Cooperative

Next, let's look at a micro-example of how much backup is needed at a small utility, right here in Vermont. When Christine Hallquist was CEO of Vermont Electric Cooperative, she spoke to my Osher grid class on May 3, 2016. Vermont Electric Cooperative is a small company that has served Vermont well for decades. You can see Hallquist's entire talk at this YouTube video (see endnotes).[136]

In her talk, Hallquist showed a chart of her system's demand on a partly cloudy day. This showed the effect of net metering solar on her system.[137] In a typical net-metering solar installation, a homeowner has a solar installation on her roof, and it sells electricity to the grid while the sun is shining. When the sun is not shining, the homeowner buys electricity from the grid. The term "net metering" refers to the fact that the homeowner's electric bill (or her payment from the utility) will be the net of how much electricity she sold to the grid and how much she bought from the grid.

Meanwhile, as Ms. Hallquist explained in her talk, the peak demand that Vermont Electric Cooperative needs to supply did not change due to net metering and installed solar. The system has to have the same availability of power to meet demand—as usual. But as Hallquist showed in her example, the power requirements for the system were no longer steady. They were spiky as solar net-metering cut in and out of the power supply, as clouds went by. Hallquist still had to have close to the same amount of nonsolar power as she had needed before the solar was installed.

This micro-example illustrates the same issues as the NBER study addressed. A grid, large or small, needs as much quick-reacting fossil capacity as it has intermittent-renewable capacity.

3. Stop-and-go driving

This is very hard to quantify but needs to be said. Running a gas turbine in an on-and-off backup mode is like running your car in stop-and-go city driving. You don't get the gas mileage in city driving that you get on the highway. Running an engine steadily is most efficient. So, using a gas turbine for renewable backup requires more gas per kWh than running the turbine steadily.

It is hard to quantify this cost because it is hard to describe exactly how spiky the situation is and how stop-and-go it is. The EPA can compare city mileage to highway mileage for a car, because the EPA has criteria for the two types of mileage. For a gas turbine backing up renewables, the amount of "stop and go" depends on circumstances.

However, there is no question that some of the renewable advantage of "clean energy" is offset by extra gas burned inefficiently as backup.

4. Efficient, gas-fired combined-cycle plants aren't run efficiently

One of the most efficient power plants on earth is a Combined Cycle Gas Turbine (CCGT). When you run such a machine steadily in base-load operation, it can turn about 60% of the energy in the gas into electric energy. In contrast, a simple gas turbine will turn about 40% of the energy in the fuel into electricity.[138]

A combined cycle plant consists of two plants that operate together. The first is a gas turbine; the exhaust from that plant is hot enough to heat water for a steam-powered cycle. The combination of the gas turbine with a steam generator leads to huge thermodynamic efficiency. You can get 50% more electricity from a unit of fuel with a combined-cycle plant than you can obtain with a simple gas-turbine cycle.

However, the second part of the cycle is usually a steam cycle, and it does not respond quickly. To back up renewables, the fast-acting part of the plant will be utilized, and the steam cycle may or may not be used. In other words, the plant will be operating at 40% efficiency instead of 60% (its efficiency with the steam cycle). Using a combined-cycle plant in a mode that is optimized for flexibility will use considerably more fuel for the same output of kWh.

5. Emissions can increase with renewable increases

Renewables are supposed to "clean up" a grid. It is certainly true that a photovoltaic installation will make no emissions as it makes electricity, and it may displace the need for a gas-fired plant to operate. In this simple example, emissions decrease with the introduction of renewables.

However, the real world is not always that simple. Emissions can also increase with the increased use of renewables. Backing up renewables can cause inefficient operation of fossil plants, leading to an increase of emissions on the grid. In other words, as renewables increase, emissions can also increase. Two studies show this effect, though both are a bit ambiguous. More renewables can mean lower emissions ... or higher emissions, as the case may be.

First, look at carbon dioxide and Ireland: A study of wind (using 2014-2015 data and published in 2016) in Ireland shows that, when the fleet of CCGT (combined-cycle gas turbines) in Ireland run steadily at about 55% fuel efficiency, the fleet produces 335kg CO_2/MWh. However, when backing up wind turbines, with more starts and stops and a lower fuel efficiency, the fleet produces 500–600kg CO_2 per MWh.[139]

However, total CO_2 per MWh for the entire Irish grid went down in 2014–2015, even though the CO_2 per CCGT went up. This is because wind displaces around 20% of fossil-fuel use. However, a later report, in 2017, shows that overall CO_2 on the Irish grid increased in 2016, due to a less-windy year and lower rainfall for hydro.[140] In the future, more wind on the Irish grid may begin to decrease the CO_2 emission rate again.

This stop-and-start driving raises the emissions of the backup gas plants. There is not a simple linear relationship between use of wind on the grid and reduced CO_2.

Second, look at NOx and Duke Energy: In preparation for a request for relaxing NOx pollution requirements, a Duke Energy spokesperson said that surges of solar energy lead to increasing emissions of nitrogen oxides. This was reported in the *North State Journal* on August 14, 2019.[141] A quote from that article:

> (Kim) Crawford (Duke spokeswoman) provided measurements showing that even on sunny days — when solar power is at its maximum output — more NOx pollution is released

into the air than would occur if no solar electricity were used and natural gas were used instead.

A few days later, on August 21, an article appeared in *E&E News* asking whether solar can increase emissions.[142] The article reported that Duke issued a statement: "To say solar is causing more air pollution? That may be some faulty logic." In this article, a different Duke spokesman , Randy Wheeless, told *E&E News* that "it's just a fact that ramping ... does cause more emissions" but implied that the emissions savings compared to steady operation of fossil fuels made up for this. As far as I can tell, Wheeless did not add more data to the data presented by Crawford but instead claimed the *North State Journal* article had jumped to conclusions.

In my opinion, I think it quite probable that ramping combined-cycle gas turbines produces more NOx than running them steadily. I worked on controlling NOx, and controlling NOx emissions is probably the most delicate balancing act in the science of pollution control. Control of NOx during the combustion cycle depends heavily on the balance between oxygen and fuel and on the temperature of combustion. Control includes controlling the air entry into the combustion process and may include shooting steam into a gas turbine to lower the combustion temperature. When a plant is changing power rapidly, these balancing acts get far more difficult and far less effective. Some power plants depend on controlling NOx after the combustion cycle, by sending chimney gases through a system where the NOx is controlled by reaction with ammonia. The amount of ammonia must be carefully balanced, to avoid "ammonia slip" (excess ammonia entering the atmosphere). Controlling the amount of ammonia is also more difficult when the gas-turbine output is changing quickly.

In short, "adding more renewables" to the grid does not necessarily lead to lower emissions.

6. Ramping up for the duck curve

As I described in the chapter on the Balancing Authority, when the sun goes down and there is a lot of solar on the grid, other power plants must ramp up very quickly. This rapid-ramping "neck" of the duck curve is energy inefficient. Think of how much gas your car will burn if you speed away from a stoplight as if you were in a race, compared to how much it will burn if you are cruising down the highway at a steady clip. It's the same with a gas turbine.

7. Lack of engineering discipline

Consider lack of steady baseload plants and engineering discipline. Advocates for "renewables can do it all" are often most scornful of baseload plants. If coal or a nuclear plant shuts down, they celebrate. It doesn't matter that the coal plant may have been polluting and the nuclear plant was not polluting. The sin that both these types of plants have committed is that they are not "flexible," and we need "flexible" plants for the new grid.

Similarly, I have never heard of an all-renewable advocate complaining about a high-efficiency gas combined-cycle plant needing to be run in less-efficient single-cycle mode. As a combined-cycle unit, it is also "inflexible," which is considered to be a major problem. But why is inflexibility a major problem? Flexibility has little to do with greenhouse gases or pollution or thermal efficiency. For renewable advocates, if a plant isn't a wind turbine or a solar installation, the most important virtue such a plant can have is the "flexibility" to back up those intermittent renewables.

This attitude shows a lack of engineering discipline. In engineering practice, equipment designs are optimized for certain uses. A diesel for backup at a hospital can be a heavier piece of equipment than a diesel for a new type of truck. A plant optimized for meeting the constant day-and-night load, a plant that runs steadily, can be optimized for efficiency and reliability, while another plant can be optimized for flexibility. That is how engineering works.

8. A kludge system

All-renewable advocates are advocating for a kludge system, not a well-designed system. In their future, all plants are optimized for flexibility, because only flexible plants that can back up intermittent renewables are allowed on the grid. Any plant optimized for steady operation, whether it is a combined-cycle gas plant, a coal plant, or a nuclear plant, is considered "your grandfather's grid." The emissions don't matter. Only flexibility matters.

An inability to acknowledge different requirements for different types of plants and the inability to make the engineering decisions that meet those requirements—that is a lack of engineering discipline.

9. Lack of seasonal reliability

Clearly, spiky intermittent renewables are not reliable in any realistic sense. Nevertheless, as in the "Spikiness" section above, most people look at renewable reliability in terms of daily output. Renewable advocates can therefore make hopeful statements about how new types of batteries will solve the problems. However, there is a bigger problem of seasonal reliability.

In climates other than the tropics, there is considerably less sun in winter. In many areas, including Vermont, sometimes solar panels are covered in snow. There is no reasonable type of battery that can be charged up on the summer solstice and provide power for the winter solstice. The only seasonal backup for renewables is fossil fuels (and whatever hydro is available).

Similarly, there is more wind at the change of seasons (spring and fall). There are more windless and very hot days in summer, and more windless and very cold days in winter. Again, it is unreasonable to consider that charging a battery in windy March will provide power to run air conditioners in August. (More about batteries in chapter 29, "Renewables and Batteries.")

10. Premature retirement of reliable power plants

I have tried to describe the direct effect of renewables on the grid and not get too deeply into policy choices. Nevertheless, it must be said that, when a reliable baseload power plant has to compete with a wind turbine that gets outside funding, the baseload power plant will often find itself paying to put power on the grid, since the wind turbine can bid in negative. When a generator bids negative into the energy market, it offers to pay the grid to take its power.

Since the wind turbines get outside funding, they can bid in at zero or negative. Their bids lower the clearing price for energy in the auction and can even drive the clearing price negative. When clearing prices go negative, all suppliers (including very reliable suppliers such as nuclear plants and combined-cycle gas turbines) must pay to put their power on the grid. This erodes the income of steady suppliers, which rely on being paid for kWhs. When prices go negative too often, these plants can be forced out of business.

Despite negative prices for energy, the overall consumer price does not go down. This is described more fully in later sections, such as the chapter "RECs Are Us." In this section, we are talking about the effects of renewables on the grid itself. Negative pricing is caused by subsidized renewables, and this policy choice drives stable plants to retire. Such retirements make the grid more vulnerable to disruption. Negative pricing is an important effect of the role of renewables on the RTO grids.

Renewables and electricity quality

In the section above, I have written about simply getting enough intermittent renewables or batteries to provide energy to the grid. I have said nothing about the quality of electricity that they provide. The electricity quality can also be a problem. As noted in chapter 5 about the Balancing Authority, many things have to be balanced in real time for a stable grid. The Balancing Authority balances energy supply, energy use, cycle frequency, and VARs (the imaginary part of the power cycle). Renewables can affect some of these issues.

Solar, wind, and batteries depend on inverters to change the direct current that they produce into the alternating current that the grid uses. The grid runs on alternating current (AC): 60 cycles per second in the US, 50 cycles per second in many other countries. The direct current from solar, wind, and batteries have to be converted to AC before it can be added to the grid.

A wind turbine is rotating electric machinery, so it actually starts out by making AC. However, the speed of the rotation is determined by the wind speed. The current cannot be added directly to lines, because the number of cycles per second depends on the wind speed, not on the requirements of the grid. Therefore, wind-turbine AC is customarily converted to direct current (DC), and then the DC is re-converted to AC at the proper number of cycles per second for the grid. Batteries and photovoltaics make DC naturally, without conversion from AC to DC. Their power must be converted to AC to be put on the grid, but they require only that one conversion cycle.

Energy from an inverter can be matched to the grid current, but it takes some action. Harmonics (overtones of the cycles per second) need to be matched, and VARs must be matched. These are pretty much solvable problems (more solvable than spikiness), but sometimes renewables are connected without the investment that is needed to match the grid constraints effectively.

Solvable issues

There are other physical issues about connecting renewables. These are often the same issues that power plants have when they are planning to connect. The issue with renewables is "Who pays for these upgrades?"

- VARS and synchronous condensers: As I described in chapter 5 on the Balancing Authority, some renewables need to have synchronous condensers added, in order to match their VARs with the needs of the grid. The owner of Kingdom Community Wind in Vermont first connected to the grid, and later they had to invest $10 million for synchronous condensers. This was described in the earlier chapter on the Balancing Authority and also in a *Burlington Free Press* article, "Can Wind Power Plug and Play with the Grid?"[143]
- Ground faults: As more power is added to the grid, transmission ground fault overvoltage (TGFOV) can occur. Under fault conditions, such as a tree falling on a power line, TGFOV can lead to serious damage to the grid, including damage to substations and transformers. When large power stations are added to the grid, the power plant developer must pay for upgrades to the substations. When homeowners add solar, the question becomes "Who should pay for the upgrade?" The homeowner whose proposed solar installation would "tip" the substation into a vulnerable state? Should that homeowner pay the $75,000 for the substation upgrade? But what about all the homeowner's friends, who put their solar panels in earlier and didn't have to pay the costs? Or should everyone on the grid pay for the upgrade, though only the customers with solar installations will make money by connecting to the grid? (That is why power plants have traditionally paid for the substation

upgrades when they connect to the grid.) Some of the issues are described in this article in *VTDigger:* "Proposed solar fee raises questions about who pays for grid upgrades."[144]

Unlike the reliability, spikiness, and seasonal reliability issues described earlier, VAR problems and ground faults have well-known physical solutions. The major question is: Who will pay the costs? When the plant owner (wind farm, rooftop solar) pays, the plant owner must cover these costs in his prices. This makes renewable energy more expensive. When the costs are spread over the whole grid, paid by all the customers, the cost of the electricity will go higher, but it will not be an easy calculation to ascribe the higher costs to the presence of renewables.

Clearly, renewables developers would prefer the second solution. From the *VTDigger* article about solar connections:

> [All Earth Renewables, a solar company, argued that] the costs associated with TGFOV upgrades "should be borne by ratepayers—the collective beneficiaries of Vermont's renewable energy laws and policies" rather than by fees on solar projects.

Daily spikiness, seasonal reliability, and the effects of inverters (needed for wind, solar, and batteries) make it difficult to integrate renewables on the grid. The more renewables, the more difficult. When more power is spiky and making harmonics, when VARs become harder to manage, then managing the grid becomes more and more difficult. The reliability of our grid would be in danger ... except for the fact that such a renewable grid would require 100% backup by quick-acting fossil plants, which could step into the breach when things get too bad.

Nevertheless, despite the reality of laws of nature—the sun doesn't always shine, the wind doesn't always blow, inverters don't make VARs for the grid—legislators make other laws saying their state grid must be 100% renewable. The laws of nature are not repealed by these renewable-mandate laws, and yet the laws are passed. Renewable-mandate laws have unrealistic plans for renewables (to put it mildly). They will not succeed in building grids that are 100% renewable. However, such laws *will* succeed in making the grid more fragile and more expensive.

So far, I have attempted to show the problems with planning to use massive percentages of intermittent renewables on the grid. Can batteries solve these problems?

CHAPTER 29

RENEWABLES AND BATTERIES

The hope of batteries

As I described in the section on the Game of Peaks (chapter 25), the peak shaving by batteries in summer is about transferring transmission costs to other utilities. Peak shaving is not about running the grid on batteries.

Sometimes batteries can be helpful. For example, in Australia, which has a great deal of wind power, Tesla has recently installed a 129 MW battery, which can supply 100 MWh of electricity.[145] I believe this system is charged by excess wind energy when wind is available, and then it is released when the wind dies down. If it were charged by fossil fuels, it would be useless: using the gas turbines directly as backup would be more efficient than charging a battery with fossil and then using the battery as backup. Capturing wind energy with a battery is much more reasonable.

Despite the hype, however, we have to notice that 100 MWh is quite a small amount when we are thinking at grid scale. Consider that Vermont Yankee nuclear plant was too small to be cost-effective, and it made 620 MW every hour, for months at a time. The Tesla battery can supply only 100 MW for one hour.

Let us say that the local grid is 5,000 MW at its peak. That would be quite small: the Northeast Grid is up to 24,000 MW at peak. For a 5,000 MW grid, you would need 50 of the Australian Tesla battery installations to store just one hour of energy.

Say the grid was heavily renewables, and you want battery backup instead of using fossil fuel. First, you would need more than an extra 5,000 MWh of renewables to charge the batteries. (There is some loss in going back and forth between the DC of batteries and the AC of the grid.) Now your battery is fully charged, but it would be able to supply the grid with backup for only one hour. If you had turned on a gas turbine for backup, it would run until you turned it off. In other words, you could run it until the sun rose or the wind picked up again.

Compared to a power plant, batteries are not very scalable to grid level.

Batteries, people, and resources

Batteries are not a source of energy. They can store energy, which is useful, but some other system needs to make the energy. Let's look at a very simple example: solar and batteries. Let us say that a homeowner puts enough solar on her roof to supply all her home's power at a time of peak demand, say a hot summer afternoon. This means the solar cells produce some extra energy at some of the times when she does not use so much power. If she wants to supply all the power to her house, day and night, she will need more solar cells on her roof in order to charge up the batteries. As a matter of fact, she will run into the same seasonal effects that challenge the grid use of solar: can she have enough solar cells charging batteries in June to supply her with electricity on a dark evening in December?

As described above, even the biggest utility battery packs are not really grid scale. Clearly, 100 MW for one hour won't do much for a grid that requires 10,000 MW baseload all night. On the other hand, a big battery pack is not the only option. A utility can arrange for individual homes to have batteries.

For example, Green Mountain Power has been renting Tesla Powerwall batteries to some customers, at a price far below what would be the customers' cost for buying them. Green Mountain Power has invested $15.5 million in the batteries for 1,100 customers and rents the batteries for $15 a month. (Some customers have two batteries.) Compared to buying batteries, the rental program is a good deal for many customers. Many battery customers have been able to use the batteries to maintain their household electricity during power outages.

However, when you get right down to it, those rented batteries belong to Green Mountain Power, and the utility plays the Game of Peaks with them. A Tesla algorithm decides when to drain the batteries and use them to reduce demand on the grid, in the hopes of paying a smaller portion of grid transmission costs.

Utility control of the batteries can lead to unexpected results for the customers. One pair of homeowners were surprised to be experiencing a power outage, since they were paying $30 a month to rent two Powerwalls. However, the utility had drained their batteries shortly before their power was interrupted by a car crash into utility poles. As the homeowner said: "We have no control over it whatsoever." In my opinion, the homeowner's statement was a bit extreme. The homeowners do have some control, but the utility has first call on the battery power. *Seven Days*, a Vermont weekly newspaper, has the story on this.[146]

Switching for a moment from the homeowner's issues to grid issues, several people who are knowledgeable about the grid (but did not want to be quoted) warned me about batteries. They said that if I even wrote about "battery backup for the grid" in my book, people in the utility industry would laugh at me. Everyone in the industry knows batteries won't scale to grid-backup level. Nevertheless, the folklore of batteries is very strong. I constantly see comments on blog posts and letters to the editor claiming that "we don't need (nuclear, a new transmission line ... whatever) as long as we have wind turbines, solar, and batteries."

But that is the folklore of batteries. People who are actually trying to figure out how to put more renewables on the grid seldom claim batteries are a significant portion of the answer for the intermittency problems of renewables. Mark Z. Jacobson of Stanford (Wind Water Solar) makes an assumption that there is more hydro available than is truly available, but he does not claim that batteries will make the difference.

An article by Nestor Sepulveda, Jesse Jenkins, et al. discussed the path to deep decarbonization. ("Deep decarbonization" is a shorthand way to describe a very-low-emissions grid. It can also mean other ways of lowering greenhouse-gas emissions). Both of the lead authors on this paper were at MIT at the time.

The Sepulveda paper describes how they modeled more than 900 scenarios.[147] Their conclusion was that using only variable renewables plus energy storage would lead to wasteful overbuilding, with curtailment wasting huge amounts of renewable energy. However, if "firm low-carbon technologies," such as nuclear and fossil with carbon capture and storage, were added to the mix, this would make deep decarbonization far easier and more cost effective.

What do I mean by saying the Sepulveda paper described "wasteful overbuilding" and "huge amounts"? Those are my words, not from the paper, and I need to back up my statements with a few quotes from the paper itself.

Overbuilding

To quote the paper: "For zero-emissions cases without firm resources, the total required installed generation and storage-power capacity in each system would be five to eight times the peak system demand, compared with 1.3 to 2.6 times peak demand when firm resources are available."

In other words, if only variable renewables and storage were available, generation and storage-installed capacity would have to be five to eight times the peak-systems demand. Such a system would need reserve margins of 400% to 700% of peak demand. In contrast, on our current national grid, reserve margins of around 15% of peak demand are common.[148]

NERC (the North American Electricity Reliability Council) lists "reference margin levels" of 10–20% in its 2019–2020 Winter Reliability Assessment, though most areas actually have higher levels.[149] NERC describes installed-capacity margin numbers. However, there are other factors that affect actual reliability, such as fuel-delivery issues. Installed capacity margins are high (70%) for winter in New England, according to the NERC assessment, but NERC notes that New England is still vulnerable to fuel-delivery problems. In other words, installed-capacity margins can be high but are not the full story. However, an 8% installed-capacity margin (as has happened recently during the summer in Texas) is tight and difficult to manage.[150]

Comparing NERC numbers with actual availability is a bit of an apples-to-oranges comparison. The installed-capacity-margin apples go as low as 8% of peak demand (installed capacity in Texas), and the oranges go as high as 70% of peak demand (installed capacity in New England, without concern for fuel delivery). However, the scenario described in the Sepulveda paper is quite different. That scenario includes renewables and storage only; it would require installed generation and storage capacity of five to eight times peak demand. This would be an unbelievably wasteful way to run a grid.

Wastage by curtailment

With only wind, solar, and storage, wind and solar would have to be overbuilt (five to eight times peak system demand) to provide enough storage. However as Sepulveda writes: "... during periods of abundant wind and solar insolation, [it will not be] cost-effective to build enough energy-storage capacity to accommodate all of this surplus ... the amount of available wind and solar output that would be wasted due to curtailment in VRE-dominated (variable renewable-dominated) scenarios would be sufficient to supply 60%–130% of total annual electricity demand ..."

In other words, after overbuilding by five to eight times, we would waste around 100% of a year's supply of electricity by curtailment ("Turn off your wind turbine—we can't use the energy now").

People who work on renewable penetration do not expect batteries to save the grid. One reason is careful modeling of battery use, as was done by Sepulveda et al. Another reason is that batteries are resource intensive.

Resource-intensive batteries

Ripudaman Malhotra, former Associate Director of the Energy and Environment Center at SRI International, is co-author of the book *A Cubic Mile of Oil.*[151] In a blog post, Malhotra calculated the amount of lithium needed for backing up a significant portion of the grid. He started his calculation by looking at what would be needed to back up one large modern coal plant. Malhotra calculates that providing 100 hours of backup for a single massive (1000 MW) coal plant would require 32,000 tons of lithium. In 2018, the global production of lithium was 62,000 tons.[152]

In short, it would take more than half a year's worldwide production of lithium to back up a single large coal plant for 100 hours.

In contrast, nickel-iron batteries use nickel and iron, which are not in tight supply. However, nickel-iron batteries need to be charged with about a third more electricity than they are able to deliver at discharge. In other words, while these batteries are not as resource intensive for the battery itself, they are resource intensive in terms of the power that they require. They lose charge at the rate of about 20% per month, whereas lithium batteries lose charge at 2% per month.[153] While the material to produce nickel-iron batteries is abundant, the number of batteries required would be much greater than the number of lithium batteries needed.

Without a Manhattan-project level of increased lithium mining, lithium-battery storage can shave peaks, but it can't provide the necessary backup for intermittent renewables. Nickel-iron batteries are wasteful in other ways. Neither type of battery is likely to be useful for backing up an entire grid. Battery storage can't provide the necessary backup for intermittent renewables.

How batteries can help the grid

All methods of energy production and storage have their strengths and their drawbacks. I admit to having heard too often about how grid-scale batteries will enable a grid with 100% variable renewables. So, I get crabby about the battery folklorists: "We don't need your nuclear, your new transmission lines, etc." However, batteries can be useful on the grid.

- Batteries can back up electricity use in homes. Very few people would buy them for this purpose, due to the cost, but when utilities such as Green Mountain Power underwrite the costs, people can use them effectively.

- In an RTO area, "peak shaving" by batteries can transfer the costs of transmission to another utility.
- Batteries come online very quickly, and their fast-ramping abilities can help maintain grid frequency on a second-by-second basis.
- When variable renewables have gone offline, and gas-fired plants must ramp up quickly to supply replacement power, a battery can supply a few minutes of power, and the gas-fired plant can ramp up more slowly. A slow ramp rate uses less fuel.

As usual, the problem is not with renewables or with batteries. The problem is that people aren't planning for their use or how they might be most useful. Renewables and batteries are overhyped and are beginning to be overbuilt. Both can be helpful to the grid. Even together, they cannot be the grid.

CHAPTER 30

RENEWABLE POLICIES

In chapter 15, "Selling kWh Is a Losing Game," I showed that a power plant selling kWh loses money. Yes, it is true that everybody uses kWh, and everybody gets billed for kWh. Yes, it is true that supplying kWh to users is the purpose of the grid.

However, on an RTO grid, a power plant can't make a living by just selling kWh. The power plants need subsidies to compete with other power plants that also receive subsidies. It's an endless battle for subsidy payments. Providing kWh to the grid is close to irrelevant.

Of course, subsidies always favor some plants and hurt other types of plants. However, in the RTO areas, the overlapping thickets of subsidies and tariffs and exceptions have given rise to the idea that "We will add more auctions and regulations until we get it right." This philosophy substitutes for a market.

First, we will review the role of subsidies. Then we will watch FERC attempt to keep the lights on by making a rule for some generators and some competitions: set-your-price-as-if-you-didn't-receive-a subsidy. The Minimum Offer Price Rule (MOPR) was designed to prevent heavily subsidized plants from completely dominating the market. Then we will watch ISO-NE follow up with its own rule, Competitive Auctions with Sponsored Policy Resources (CASPR), which tries to help plants that might have been hurt by MOPR. The CASPR rule was cleverly designed to help some plants in some competitions but hopefully not affect other aspects of related competitions. ("We will add more regulations until we get it right.") We will also take a side trip into other RTOs dealing with renewable subsidies.

As all this unfolds, interested parties will proclaim that the latest iteration of rules is "breaking the market." In general, these proclamations come from groups whose financial interests will be harmed by the latest set of rules.

RTOS are not the only complication

In planning this book, I felt it was important to show the complicated situations that are caused by the rules in the RTO areas. I didn't want to start by describing issues around renewables and carbon dioxide and nuclear.

- Carbon dioxide, renewables, and the role of nuclear energy are hot-button issues for many people.
- Is the rise of carbon dioxide a huge problem or a "lefty plot"?
- Can renewables power the U.S.? Can they power the world?
- How much should renewables be supported, if they raise the cost of power for poor people? Do renewables raise the cost of power?
- Meanwhile, nuclear is low-carbon, but are people willing to use it? (Germany's Energiewende is more concerned with closing nuclear than with lowering Germany's carbon footprint.)

- Should nuclear plants be subsidized, as are many renewable plants?
- Or is subsidizing nuclear plants a travesty and a crime?
- Is subsidizing renewable installations a travesty and a crime?

In terms of carbon and renewables and nuclear, emotions run high, and beliefs are firmly entrenched. If I had started my description of the grid by discussing renewables on the grid, and how they operate and how they are paid for, readers might not see the more general RTO grid-governance issues at all. Many of the readers would be caught in their emotional reactions to the renewable-carbon-nuclear issues.

That is why I started with how much trouble an RTO had with the comparatively simple problem of keeping the lights on during the winter in New England. But it is time, now, to bite the bullet, take the bull by the horns, and pull out every cliché in the book. In other words, it is time to dare to talk about renewables and subsidies on the grid.

(Warning. This section will lead us into another acronym jungle.)

Federal subsidies and requirements

Many of the renewables issues are caused by state policies. However, there are several federal programs set up to support renewables.

The major federal program is the Production Tax Credit (and its sibling, the Investment Tax Credit). These provide subsidies for renewable projects. An Investment Tax Credit gives tax credits for investment: that is, if you build the project (for example, a wind farm), you get a credit against federal taxation. Projects begun in 2015 and 2016 received a credit of 30% of the investment against federal taxation. This credit began ramping down in 2017 and is due to end in 2020.

The more commonly used tax credit is the Production Tax Credit. In 2015, this tax credit was 2.4 cents per kWh, and it is also phasing down. When this tax credit was fully in force, a renewable developer would receive a 2.4 cent credit against federal tax liability for each kWh he put onto the grid. In other words, this is a tax credit for production of renewable kWh, rather than for investment in a renewable facility. The tax credit is scheduled to end in 2020.

In 2014, as reported in *U.S. News and World Report*, Warren Buffet told an audience in Omaha, Nebraska, that subsidies are the only reason to build wind farms.[154]

"I will do anything that is basically covered by the law to reduce Berkshire's tax rate," Buffet told an audience in Omaha, Nebraska, recently. "For example, on wind energy, we get a tax credit if we build a lot of wind farms. That's the only reason to build them. They don't make sense without the tax credit."

As an example of this credit, say Buffet owns a 2 MW wind turbine, which has 30% capacity factor. There are 8,760 hours in a year, so the wind turbine would produce 5,256,000 kWh of energy. Each kWh would earn approximately 2 cents production tax credit (the amount varies over time) so the wind turbine would produce a $105,120 credit. That is the amount of the tax credit: that is, the amount that the company who owns the wind farm can deduct from its federal tax bill.

For a large, profitable company, such a deduction is more valuable than simply receiving this money as income or even as profit. The company would pay taxes on its profits, but when a company lowers its tax bill, that is straightforward money in the bank. As Zach Starsia writes in a post on investment advice, production tax credits can supply "tens or hundreds of millions of dollars" to the investor, and "can represent as much as 60% of the project cost."[155]

For small wind turbines of 100 kilowatt or less, the investment tax credit is 30%, but 12% for larger wind turbines.[156] As far as I can tell, the Production Credit is the more lucrative credit for a large wind turbine, and I will continue to discuss the Production credit only.

When I say that these credits are ramping down, it is important to note that they have ramped down and reappeared several times.[157] Another thing to remember is that the production tax credits continue for years. So, the statement "Production tax credits will phase out next year" would have very little to do with what existing renewable installations will receive. The 2013 renewal of the Production Tax Credit for wind provided 2.3 cents per kWh tax credit for the first 10 years of production for a facility completed in 2013. A facility completed in 2017 would receive 20% less per kWh for the first 10 years of production, and facilities that begin construction in 2020 would not receive the production tax credit. However, there's a catch called "safe harbor," in which a developer can receive a tax credit if he has spent 5% of the cost of the project in a year in which he would still be eligible for the production tax credit. A U.S. Energy Information Administration article provides a detailed summary of the production tax credits.[158]

Every time the subsidies are due to ramp down, building renewables ramps up. The subsidies were due to end in 2012. As described in the EIA article above, that was the all-time record year for wind projects.

One close-to-home example of the rush to complete wind for tax credits shows how fierce it can get. Here in Vermont, a wind developer was in a hurry to complete his project by the end of 2012. If he wasn't connected by the end of 2012, the production tax credits for his project were due to expire.[159] He was in such a hurry that he sued his neighbors in August of 2012. They objected to the wind project, and they stood on their own property, near the property line, and did not get out of the way of his blasting. But he had to blast in order to move quickly on his project. The developer sued the neighbors for standing on their own property. The judge ruled for the developer, and the neighbors retreated.[160]

For the developer, this story had a happy ending. The wind project was completed and connected to the grid by the end of 2012, the dates when the tax credits were due to expire.[161]

Actually, the tax credits were extended past 2012. The history of the federal tax credits is a zombie history. The tax credits are always supposed to die. A recent (July 2019) article in Reuters illustrates how the tax credits are revived.[162]

So far, these tax credits have always been renewed.

States

The main production and investment tax credits are federal, but most renewable subsidies are state level. (Some states also use tax credits to encourage renewable energy.)

A state can enact a renewable portfolio standard, and then the utilities in the state must buy a certain percentage of their electricity from renewable sources. A state can enact a zero-emissions credit, and then utilities in the state will be required to buy zero-emission electricity, which generally includes renewables and nuclear. The utilities will usually be mandated to pay some extra fee to the zero-emission suppliers.

Similarly, ISO-NE may tie itself in knots, making rules to encourage dual-fired power plants to keep oil on-site for winter. The ISO-NE rules must meet FERC requirements to be fuel neutral. (See the chapters on Jump Ball). While this is going on at ISO-NE, a state can make a rule that, within that state, power plants must meet strict emissions limitations. In the December 2018-January 2019 cold snap, Massachusetts plants were nearing "hard limits" on the amount of carbon dioxide they could emit, which could have forced them to stop burning oil. These rules constraining oil are somewhat in opposition to ISO-NE's attempts to keep oil available at dual-fired plants for grid stability in winter, but a state can make such rules. Similarly, a state can set up liberal payments for net metering, and a great deal of net-metered solar power will be installed. Then ISO-NE will have to dispatch traditional plants quickly at dusk, as the sun goes down, and the rooftop solar goes down with it.

There is a balance of power between the RTOs and the states. In terms of renewables, the states have the most freedom of action. The RTOs just follow along, trying to keep the lights on at what they hope will be low costs.

The overall influence of these subsidies and credits is shown in figure 8, "Revenue streams of different types of plants," in chapter 15, "Selling kWh Is a Losing Game."

Who helps whom?

In New England, NEPOOL wants to help straighten out the mess of state laws and ISO-NE mandates. NEPOOL started the IMAPP process (Integrating Markets and Public Policy). But what is the purpose of this process? And yes, I am always suspicious about a NEPOOL initiative.

According to the NEPOOL website, the purpose of the IMAPP process is to "explore potential changes to the wholesale power markets that could be implemented to advance state public-policy objectives in New England."[163] NEPOOL seems to be saying that FERC 1000 wasn't enough to help the states meet their renewable objectives. The states want more help.

In my opinion, a certain amount of dramatic tension is a good thing. Our constitution has a "balance of powers" between the executive, legislative, and judicial branches of government. In theory, no one branch can rule the roost. Similarly, in my opinion, an RTO or PUC that is pushing for reliability and low cost can be a us against excessive state mandates. State mandates migh many renewables too fast: this could raise grid prices grid reliability.

The idea that ISO-NE will work to "advance state policy objectives" in New England appalls me. I feel like screaming at ISO-NE: "Keep your reliability mandate going! Don't give up the ship. We need you."

I think that many of the members of the NEPOOL Participants Committee would disagree with me. From their point of view, ISO-NE just keeps getting in the way.

As noted in chapter 15, "Selling kWh Is a Losing Game," in many cases, renewables don't have to make any money by actually selling their energy to the grid. They make money by selling RECs (Renewable Energy Certificates) and by receiving production tax credits. Renewables can pay the grid to take their power (negative pricing) and still come out ahead financially.

Both RECs and production tax credits are paid by the kWh. If the renewables don't provide kWh to the grid, they don't get paid those extra compensations. So, their finances depend on selling power to the grid (at a loss) and making their money with the non-grid compensation.

This has several effects. Due to the way grid pricing works, if there are low prices on the grid for one set of units, this will generally lower the clearing price and, therefore, the compensation for all the other units. That is, zero-cost energy bids will lower the grid prices for selling kWh. This is widely trumpeted by renewable advocates as "Due to renewables, prices are going down." But there is a catch. They mean "wholesale prices on the grid kWh auctions are going down." Prices to the consumer are going up.

Grid prices go down. User prices go up.

THE GRID PRICE per kWh is a fallacious way of accounting for renewable costs because the grid price for kWh does not show the entire picture of what the customer pays.

Let's look at a generator that is selling renewable kWh on the kWh ction. Say that this is a wind farm and is bidding into the auction at zero cents per kWh. Due to the auction method, the price for kWh on the grid will be lower, due to the presence of the wind farm. Yes, the clearing price at the auction will probably be lower.

However, the wind farm also expects to sell RECs (Renewable Energy Certificates) as well as kWh. Some utility will have to buy those RECs to meet a renewable portfolio standard. The RECs will then be part of that other utility's overhead, and, therefore, a ratepayer will pay for the RECs. One ratepayer is paying the grid clearing price for the wind kWh, and another ratepayer is paying for a wind REC through his distribution utility.

Like many greenwashing strategies that utilities use (such as the Game of Peaks), RECs don't save money. They just move it around from one ratepayer to another. The next chapter looks more closely at those RECs.

CHAPTER 31

RECs ARE US

RECs and their sales

EVERY NOW AND AGAIN, we will see an announcement that some company is doing its part for the environment by using "100% renewable" electricity. This is supposed to make us imagine a big industrial facility surrounded by wind turbines and solar panels. Well, no. That's not the right image. Better to think about an accountant.

When a renewable-energy source such as a solar panel or a wind turbine makes a kWh of electricity, it also makes something invisible: a REC (renewable energy certificate) worth one kWh of renewable credit. It sells the electricity to the grid, and it sells the REC somewhere else.

Usually, RECs are sold to a distribution utility in a state that requires utilities to use a certain percentage of renewable power. With a fistful of RECs, the utility can claim to have bought the correct amount of renewables. The RECs are the renewables, from an accounting point of view. Buying the REC gives money to a renewable facility, and that helps it be competitive with other facilities. On the other hand, RECs (pardon my opinion here) are very misleading. A utility in Connecticut can buy a REC from a wind farm in Maine, and the Connecticut utility can claim to be using renewable electricity. Meanwhile, almost all the electricity on the local grid comes from high-emissions power plants burning gas, and from low-emissions sources such as nuclear and large hydro plants. In most states, the nuclear and hydro plants are not allowed to make or sell RECs. (Remember, the whole REC thing goes by state.)

RECs don't even have to be on the same grid as the REC buyer. A Connecticut utility buying RECs from Maine is bad enough. The power from Maine is being used in Maine, or maybe in New Hampshire. Nobody can track an electron, but it is unlikely that the power made in Maine is finding its end user hundreds of miles away in Connecticut. But at least the two states are on the same grid system.

REC markets can be even broader. As an article in *EnergySage* reports:[164]

> Organizations and individuals can buy RECs from any supplier, which means that a company in Vermont can buy RECs generated in South Dakota. The company would save some money by doing so ... [but] buying more expensive Vermont RECs would be a show of support for local renewable energy generation.

This is a dramatic example (people in Vermont buying RECs from South Dakota), but the statement that "Organizations and individuals can buy RECs from any supplier" is too broad. RECs are administered by state, and the states make and change their rules. For example, out-of-state suppliers used to be able to participate in the Pennsylvania Solar REC market. In 2017, Pennsylvania restricted eligibility for that market to solar photovoltaic systems sited in Pennsylvania.[165] Still, according to the rules in many states, buying a REC does not necessarily require the buyer to be anywhere near where the renewable power is being supplied.

Oh, my. I have now mentioned solar RECs and states. It seems worthwhile to note that states have many different rules for many kinds of RECs. For example, James Bride, of Energy Tariff Experts, described the REC requirements for five New England states. This description was in a presentation he gave at the Consumer Liaison Group meeting of ISO-NE in 2015.[166] These REC requirements are driven by the states' renewable portfolio standards, and they slice the required renewables into 15 different classes, each with a different kind of REC. Massachusetts has the most classifications (six). Classifications include Solar I, Solar II, Combined Heat and Power, small hydro, waste-to-energy, and Alternative Energy Standard.

Bride then calculated the "compliance costs to load" by state, for selected time periods in 2015. Compliance costs are the extra costs imposed by meeting the state's renewable portfolio standards. The costs varied from 0.2 to 0.3 cents per kWh (Maine) to 1.1 to 1.6 cents per kWh (Massachusetts).

I appreciate that James Bride did these calculations. It's pretty easy to see the prices per kWh in the auctions on the grid, but the prices for the RECs get folded into the other overhead for utilities. It truly takes an Energy Tariff Expert ("Energy Tariff Experts" is the name of Bride's company[167]) to tease out these costs.

The ratepayers (nobody calls us "customers") pay these overhead expenses. With grid prices running at an average of 4.1 cents per kWh in New England in 2015[168] (the year of Bride's study), the price rises caused by renewable-portfolio standards are significant.

Two types of entities purchase RECs:

- Utilities that have to meet their renewable-portfolio standards for renewable purchases. This is called the "compliance market." These RECs tend to be more expensive, because utilities need to buy a certain amount.
- Businesses (mostly) that want to wave a green banner in front of their customers. This is called the "voluntary" REC market, and the RECs tend to be less expensive.

The circumstances under which a renewable-energy provider will sell into one or the other of these markets is beyond the scope of this book. Let's just say they will sell into the compliance market, if they can.

Which gets back to the question: Are RECS a shell game? If a utility in Pennsylvania buys solar RECs from Maryland, what does this mean? The solar providers are still getting some more payment, though the Pennsylvania utility is actually using electricity from gas, coal, and nuclear. The Pennsylvania utility is not using the solar power that goes with RECs.

In some ways, it is not a shell game. Renewable-energy suppliers are being supported by RECs. They get the money.

In other ways, RECs are a shell game. The user is not depending on intermittent renewables for its electricity. The user can claim to be "using renewable energy," but that is about accounting, not about the energy they are actually using.

I guess "shell game" is in the eye of the beholder.

Nationwide costs for renewables

RECs ARE ONLY PART OF the costs of renewables. Estimating the costs of renewables, nationwide, is practically an impossible task. There are so many types of costs, and they vary by jurisdiction. I will list a few here.

Subsidy costs—RECs: Probably, the first step would be looking at the costs for RECs. As noted above, some states have several types of RECs, with different costs for the different types.

Subsidy costs—net metering: Also, net-metered solar usually means paying solar owners retail prices for wholesale power. The net-metered solar owner gets paid, say, 14 cents per kWh (his retail cost) for supplying electricity that the grid could buy for 5 cents from a fossil or nuclear plant. The difference between wholesale and retail prices for power is supposed to cover transmission and distribution costs. It also covers other costs, such as billing, administration, and crews to restore power after an outage. The net-metered owner gets paid the retail price, while supplying only wholesale-type power. The net-metered owner doesn't contribute to the transmission and distribution costs.

As you might expect, these net-metering payments raise the retail price for everyone else. Some areas are now charging "connection fees" to net-metering owners, to cover at least some of the costs of maintaining the grid. In Vermont, some early proponents of net metering are now concerned about their overly generous plans. Tony Klein is the former chair of the Vermont House Natural Resources and Energy Committee. He was an early proponent of net metering. Now Klein is concerned about the effect it is having on smaller utilities, perhaps driving some to the brink of bankruptcy. Ken Nolan is the general manager of Vermont Public Power Supply Authority, which represents small rural cooperatives. Nolan notes, that, in a small utility, the subsidies required for a single, large net-metering project could lead to a rate increase for all other customers.[169]

Subsidy costs—curtailment: In some areas, wind turbines get paid if they cannot get online at the times that they are available. Only renewables get this type of payment. For example, in Washington State, the Bonneville Power Administration used to ask thermal generators (such as coal, oil, and gas) to go offline when the rivers were full and hydro plants could make lots of inexpensive power. Bonneville Power is both the authority running most of the hydro plants for the Northwest *and* the Balancing Authority for that area. Not being dispatched was okay with thermal plants, because they didn't burn fuel when they weren't dispatched.

However, as wind turbines became a bigger presence in the Bonneville region, their owners complained to FERC about not being dispatched when the rivers were running high. When a wind turbine is able to run (the wind is blowing) but it is not dispatched, it loses the money it could have made in production tax credits and selling RECs. FERC agreed with the wind turbines, and now Bonneville Power has to compensate the wind farms for lost revenue if it does not dispatch them when the wind is blowing. The wind turbines must be paid when they are curtailed.[170] This ruling is only for the Pacific Northwest: wind turbines in other areas are paid or not paid when curtailed, depending on the jurisdiction.

It would take a nationwide research project to estimate what consumers pay for curtailing wind, in each area and each month.

Redundancy costs: Meanwhile, as Christine Hallquist noted in her talk about Vermont Electric Cooperative, she has to have as much backup capacity on the system with renewables as she needed without renewables. The NBER study, a worldwide review, determined that a grid needs 1.14 MW of installed fossil capacity for each MW of intermittent renewable capacity. (Both of these are discussed in chapter 28, "Three Issues with Renewables.")

Transmission costs: The costs of transmission are steadily going up. Some fraction of the increased costs of transmission are due to the expense of connecting far-flung renewables to the grid. However, it is almost impossible to say whether this is a large or small portion of the increased costs.

I am attempting to estimate only the renewable costs that are paid by utilities and reflected in the consumer electricity bills. There are production tax credits and other tax credits that are paid through taxes, and there are local tax breaks and subsidies for renewables, sponsored by the town or the state. In this section, I am attempting to show the difficulties of estimating the costs of renewables that are paid by the electricity customer, not the costs paid by the taxpayer. Of course, they are, generally, the same person: people both buy electricity and pay taxes.

I have heard the statement that "cheap renewables are undercutting traditional generation" far too often. The renewables are rarely, if ever, cheaper than traditional generation. But just like New England, with its six types of solar RECs in Massachusetts, it is almost impossible to figure out the costs of renewables nationwide, due to the many programs and cost shifts involved.

Starting from first principles about renewable costs (How much is spent on RECs? What are the system costs of the need for redundant backups?) is an almost-impossible job. That is a bottom-up approach that depends on identifying all the costs, in all the jurisdictions.

A recent study at the University of Chicago took a different approach. (The University of Chicago is my alma mater, so, of course, I like the study.) Instead of trying to piece out the costs bottom-up, the University of Chicago study looked at the price results of renewable portfolios. They used a systems approach. They asked a simple question:

How do renewable portfolio standards affect the customer's bill?

Renewables and consumer prices

In the Chicago study,[171] Michael Greenstone and Ishan Nath described seldom-considered features that increase costs for renewables:

> 1) The intermittent nature of renewables means that backup capacity must be added; 2) Because renewable sources take up a lot of physical space, are geographically dispersed, and are frequently located away from population centers, they require the addition of substantial transmission capacity; and 3) In mandating an increase in renewable power, base-load generation is prematurely displaced, imposing costs on ratepayers and owners of capital.

Instead of trying to track all these separate costs, Greenstone and Nath looked at the effect that renewable portfolio standards had on consumer bills. For the states that have instituted standards, they compared the consumer prices in that state before and after implementation. They found that renewable portfolio standards raise the consumer cost of power. A quote from their abstract:

> Renewable portfolio standards (RPS) are the largest and perhaps most popular climate policy in the U.S., having been enacted by 29 states and the District of Columbia. Using the most comprehensive panel data set ever compiled on program characteristics and key outcomes, we compare states that did and did not adopt RPS policies, exploiting the substantial differences in timing of adoption. The estimates indicate that, 7 years after passage of an RPS program, the required renewable share of generation is 1.8 percentage points higher and that average retail electricity prices are 1.3 cents per kWh, or 11% higher; the comparable figures for 12 years after adoption are a 4.2 percentage-point increase in renewables' share and a price increase of 2.0 cents per kWh, or 17%.

Renewable portfolio standards are a very expensive way to abate carbon. Greenstone and Nath found that, with a renewable portfolio standard, the abatement cost per metric ton of carbon was between $130 and $460. This is far higher than the cost per ton in cap-and-trade systems such as the Northeast's Regional Greenhouse Gas Initiative (about $5 per ton) or California's cap-and-trade system ($15 per ton). The authors estimated that consumers in the states with renewable portfolio standards had paid a total of $125 billion more for electricity than they would have paid without the policies.

Instead of looking at renewable pricing and renewable subsidies in heavily manipulated RTO markets, this study looked at the cost of renewable mandates to consumers. The cost to consumers reflects the system costs of renewables. It is not a pretty sight.

The University of Chicago study was nationwide in scope, and it estimated the increase in average prices for complying with renewable mandates to be around 1.3 to 2 cents per kWh. James Bride looked at "compliance costs to load" for renewable power in New England. He found these extra costs for the renewables to be 0.2 to 1.6 cents per kWh. These two studies covered different areas of the country and had different methodologies and somewhat different results. They both showed significant costs for integrating renewable energy on the grid.

The midnight sun

THE SUN MAY SHINE only during daylight hours, but a business may decide to buy solar RECs that add up to 100% of the power it uses and then advertise that it runs on 100% solar. Since the electricity and the REC are disconnected, you can use a solar REC at midnight, which is very misleading. "Look, that business runs on 100% solar. Why doesn't everyone?"

Being "100% renewable" by buying RECs gives a business the best of both worlds: bragging rights and virtue signaling for its customers *and* reliable grid power for its operation. Its electricity is "100% renewable," but the business has no worries about intermittent power or electricity storage. Great for the company, though misleading for most of the people hearing about its "100% renewable power."

Also, utilities and businesses have arbitrage possibilities due to different states having different criteria for RECs, and even several types of criteria within a state (six for Massachusetts, as we see in the Bride presentation). Vermont buys RECs from Hydro-Québec at a low price, because nobody else wants them, and then sells its own RECs to southern New England states, where they want RECs to meet their renewable portfolio standards.

This REC arbitrage cuts my personal power bill here in Vermont. Other states pay part of my costs. Why should I be against it?

I'm against it because, with complex rules and state-by-state complexities, what could have been a good idea has, in my opinion, devolved into a game, based on arbitrage and heavily driven by public-relations opportunities.

To quote Annette Smith of Vermonters for a Clean Environment: "The RECs are symptomatic of energy policies so complicated that few people understand them, including many of the politicians who voted for them."[172] Smith is active in shaping energy policy in Vermont. On the other coast of the continent, Professor Leah Stokes of UC Santa Barbara has a new book about energy policy in which she describes how "The 'fog of enactment' (the gap between actors' expectations and the policy's actual outcome) helps explain why it is so difficult to *implement* clean energy policy."[173] I would summarize their statements in this way: when the policies get too complicated, the results of such policies become hard to predict, even by the people who wrote the policies.

Looking at the whole complex system, I think the games-with-RECs are going to backfire on the renewable industry. At some point, the shell games will be exposed. The bad news isn't even that the RECs might backfire. The bad news is that complex policies and games-with-RECs do not lead to a reliable grid.

"RECs Are Us" is another road to grid fragility.

CHAPTER 32

VERMONT, THE TWICE-SOLD STATE

Meanwhile, in Vermont

THERE ARE SIX New England states, and Bride described the REC requirements and renewable portfolio standards for five of them. What about Vermont?

We don't have a renewable portfolio standard in Vermont. We encourage our utilities to buy renewable power with several programs, as described by Annette Smith in her white paper.[174]

Smith is a well-known opponent of Big Wind and is very knowledgeable about energy policies in this state. As a matter of fact, she is so knowledgeable that the Vermont Attorney General once investigated her for "unauthorized practice of law" before the Public Service Board. She was accused of providing legal advice to people about renewable-energy projects. Smith had never represented herself as a lawyer but would have happily said that she advised people on how to fight intrusive projects.[175] The public outcry against the Attorney General's investigation may have been one of the reasons that the investigation was quickly dropped.

In Vermont, utilities have many different incentives to use renewable energy, and the situation is deliberately complex. To learn more about Vermont's various programs (Renewable Energy Standards, with three tiers in the program, a Standard Offer program, Act 174 Energy planning, and so on), I recommend Annette Smith's white paper referenced above.

Basically, if a Vermont solar panel makes some power, it can sell the REC for that power out of state, where it hopes to get a good price. Meanwhile, Vermont can satisfy its own REC requirements (such as they are) by using RECs produced by hydro power from Hydro-Québec (HQ). No other state in New England recognizes the Hydro-Québec dams as "renewable energy," so Vermont can purchase these HQ RECs very cheaply. Meanwhile, it can sell its own RECs for higher prices in another state.

This is called "REC arbitrage." It turns a profit for Vermont, though the profit has been falling as REC prices decline in other states. Vermont has found a way to make money from RECs, and southern New England states have found a way to use Vermont RECs to label their own grid power (mostly natural gas) as "clean."

Personally, I should not object to RECs, because RECS sold out of state subsidize my electricity bills, here in Vermont. Why should I be concerned with something that helps my pocketbook? Still, I feel compelled to tell the sad history of my state and its virtue signaling.

Basically, for many years, Vermont did not have a renewable portfolio standard and did not require utilities to submit RECs to meet such a standard. Vermont did require or encourage utilities to use renewable energy, but not through RECs. Also, Vermont utilities are not fully separated into distribution and generation utilities. Such a separation is the usual thing in the RTO areas. Vermont is in an RTO area, and our utilities do take part in the auctions, however, our utilities can also own power plants. For example, the Kingdom Community Wind project on Lowell Mountain is owned by Green Mountain Power. Yeah, it's complicated.[176]

For many years, the lack of a renewable portfolio standard in Vermont meant that a utility could satisfy its Vermont requirement for renewable energy by pointing to the number of kWh it received from renewable sources. If it owned the renewable resource (such as Kingdom Community Wind), it could use the facility's kWh count to satisfy the in-state rules, and it could sell the RECs for those same kWh out of state. Actually, it didn't matter who owned the renewable facility. The owner could sell the renewable kWh to a Vermont distribution utility, who would claim these kWh as meeting Vermont renewable requirements. The owner could then sell the RECs for those same kWh to a state that had a renewable portfolio standard requiring RECs. One kWh equaled two kWh worth of credits for renewable energy.

Yes. People noticed.

I would like to say that my state came to its senses, decided to straighten up and fly right, that it decided not to double-book anymore, and that it had a change of heart. However, that is not what happened.

Starting around 2015, Vermont's double-booking became a controversy that spread beyond Vermont. The Federal Trade Commission warned Green Mountain Power about its Renewable Energy claims, but ultimately backed off from prosecuting the utility for fraud.[177] Meanwhile, as described in a *Utility Dive* article,[178] NextEra Energy dropped its purchases of Vermont RECS. The NextEra decision followed another decision: the Connecticut legislature had placed a ban on purchasing Vermont credits.

You can't double-sell if nobody is buying. These market pressures led the state to straighten up and stop double-selling.

Double-selling was pretty big business. *VTDigger* reported that, if Vermont lacked the ability to sell RECs to other states, Vermont electric rates could jump 6% statewide and 20% in areas like Burlington that were proud of their heavily renewable electricity mix.[179] Meanwhile, new legislation was introduced. This legislation was very complicated, and it is quite possible that even the legislators who passed the laws did not understand them (see Annette Smith's comments in the previous chapter). As far as I can tell, the laws were designed to not lose all of the $50 million that Vermont utilities received from other states for Vermont RECs.

Vermont shapes up

Vermont can meet its own renewable-energy requirements with REC equivalents from Hydro Québec. No other state in this area considers RECs based on the big Canadian dams to be valid for meeting renewable-energy requirements. Vermont does, and this means that we can still sell many of our own RECs out of state.

The new rules make Green Mountain Power's website somewhat difficult to understand.[180] Green Mountain Power is the biggest utility in Vermont. Their site shows fuel mix before and after REC sales.

Looking at their site as it appeared on December 13, 2018, it showed that, before REC sales and purchases, the GMP fuel mix was 23% big hydro (existing Vermont hydro is called out separately), 14.7% nuclear, 30% market purchases. Before any REC sales, Vermont renewables included 5% solar, 8% wind, and 5% wood. After REC sales, the mix was a 49% large hydro and less than 1% renewables.

Looking at the site again on May 7, 2020, Green Mountain Power has combined the percentages for Vermont hydro and HQ hydro. This makes the REC situation even more difficult to understand.

Basically, looking at the more-complete data, as it appeared in the 2018 website, Vermont renewable RECs had been sold out of state, and large hydro RECs (much cheaper) have been bought in their place. (See the numbers in the paragraph above, where I describe the site as of December 2018.) I admit the numbers don't completely add up, but clearly the Vermont renewables are gone (sold as RECs), and more large hydro is listed (bought as RECs).

Well, selling Vermont RECs out of state and buying RECs that nobody else wants—this is still helping my electricity bill. So, why should I care?

Supposedly, Vermont is in the very forefront of clean energy. In 2011, the Vermont Department of Public Service published a Comprehensive Energy Plan for Vermont. According to this plan, Vermont will get 90% of its energy from renewables in 2050—that is, 90% of all its energy, including heating and transportation. Now, clearly, there are some practical problems with this. I don't think that any heavy truck can run on renewables, as yet, but let's not look too deeply into the exact methodologies Vermont plans to use. Vermont's aspiration to go to 90% renewables shows that Vermont cares.[181]

Or does Vermont care? In 2012, CVPS (Central Vermont Public Service) offered "cow power" electricity. (CVPS is now owned by Green Mountain Power.) An electricity user could pay a little more for electricity and be assured that she was getting "renewable power." It was called "Cow Power" because one of the sources of the renewable energy would be methane from digesters at dairy farms. Did Vermonters rush to sign up for Cow Power? Not really. Two percent of CVPS customers (3,300 out of 160,000) signed up.[182]

Currently, Green Mountain Power has a Cow Power program at four cents per kWh extra fee.[183] The earlier CVPS program was limited by the availability of cows, although I was under the impression (at the time) that the availability of customers for Cow Power was also an issue. The current GMP program is not limited by the number of farmers available to sign up for digesters: the program can also buy RECs from other states or add money to a Renewable Development Fund. The information about the number of GMP Cow Power sign-ups is not on the GMP website, but spokeswoman Kristin Kelly shared this information with me in an email.[184] GMP has 2,016 customers enrolled in Cow Power. According to their website, GMP has approximately 265,000 customers in Vermont.[185] This means that less than 1% of its customers have signed up to pay more for local renewable electricity. In fairness, Ms. Kelly reassured me that many GMP customers take part in other emissions-reduction programs supported by GMP. (My impression is that many of these programs save money for the lucky end-user, instead of costing money.) Looking at the bottom line, we see that very few GMP customers have signed up to pay more for renewable electricity.

Renewables versus Yankee thrift

IN OTHER WORDS, DESPITE the 90% renewables plan for the state, when people in Vermont have a choice of paying more or paying less for electricity, they generally choose to pay less. When Vermont has a choice of selling RECs or keeping RECs for renewable power, it sells the RECs to other states and buys (otherwise worthless) RECs from Hydro-Québec.

In short, when the electricity choice is between Green and Thrifty, Vermonters generally choose Thrifty.

It's still a Yankee state up here, despite all the immigrants from big cities, the high-priced summer homes, and so forth. Buying an organic apple for more money: that is something Vermonters will do. The organic apple is considered to be a better apple. Buying fancy electricity: heck, it's the same electricity everyone gets. Why pay more? Especially when we can buy RECs from Canada.

Vermont is no longer quite the twice-sold state that it used to be, back in 2015 when the Federal Trade Commission was looking into our REC practices. However, Vermont's electricity rates bounce around at being about the fifth-highest state in the country. This isn't a state where people want their already-high electricity bills to go higher.

Virtue signaling by announcing a plan for 90% renewables is one thing. Digging deeper into your very own pocket is quite something else.

I live in the twice-sold state.

CHAPTER 33

THE PURPOSE OF RENEWABLES

Zero emissions

In the last few chapters, we have looked at some aspects of renewable energy, specifically, the buying and selling of RECs in various states. Here's a deeper question, however: What do people want when they say they want renewable energy? What is it about renewable energy that is so attractive to many people? Why do states and federal governments arrange subsidies to encourage renewable energy?

In my opinion, the answer is fairly simple: people want to stop using fossil fuels, and they want clean air. To me, "clean air" means that the plant does not emit SOx (sulfur dioxide), NOx (nitrogen oxides), or particulate. For me, and for many people, it also means that the power plant does not emit carbon dioxide. (Yes, I think man-made carbon dioxide is a problem, but that is not what this book is about. It's also completely okay if you disagree with me on this one. This book is about the grid, not about the climate debate.)

A plant that does not emit SOx, NOx, particulate, or carbon dioxide is often described as a "zero-emissions" plant.

Nuclear plants don't use fossil fuels and don't emit combustion products into the air. As a matter of fact, nuclear plants make the majority of the zero-emission electricity in this country: 55% is nuclear, 20% is hydro, and 19% is wind.[186] Counting all types of electricity generation (coal, gas, oil, nuclear, wind, hydro, etc.), nuclear makes 20% of the electricity in the United States.

It would be nice to say that people have begun to notice the value of nuclear, due to its clean-energy abilities. Actually, various states began to notice something more immediate: they wouldn't meet their low-emissions goals unless they kept their nuclear plants running.

Renewable installations get subsidies in order to compete in the market. With low prices for natural gas, and with nuclear plants beginning to close for financial reasons, many states decided that they needed something like REC for nuclear plants, in order to keep them financially viable. Renewable plants get a considerable percentage of their funding from production tax credits and RECs, while nuclear plants are stuck with selling kWh to earn their money. To keep nuclear plants financially viable, the ZEC (Zero Emission Credit) was born.

I was there for the first birth of a ZEC. I can take very little credit for the ZEC being put into place, despite my long-term pro-nuclear advocacy. But I was there to celebrate! I was in Albany, New York, when it happened. Here's a firsthand report.

The New York rally

On August 1, 2016, I was part of a rally in Albany, New York. The rally supported the New York Clean Energy Standard, which is a kind of Zero Emission Credit. Later in the day, the New York Public

Service Board enacted the low-emission-credit rule. At that point, we were rallying in celebration.

It was overwhelmingly wonderful to be there.

This description is based on my blog post about the Albany event.[187]

My visit to Albany started the night before the rally with a dinner for nuclear activists from all over the country: California, Ohio, New York, Vermont, New Hampshire, Virginia. Rod Adams includes pictures of the dinner at his blog post: "Fighting climate change with best available tools."[188] Several people at the dinner knew me through my blog posts, and one man took a "selfie" with me. I was not only among friends, but I was a minor celebrity. (Believe me, that never blunts the feelings of joy.)

The next morning, we gathered for a rally before the Public Service Commission meeting. Our group met in a ground-floor corridor of the building in which the official hearing would take place. The hearing would be on the 19th floor.

In our corridor meeting, several people addressed the group. I spoke about the consequences of closing Vermont Yankee and why we have to avoid closing nuclear plants. Eric Meyer led us in a rousing rendition of "The Battle Hymn of the Atom" (The Truth Goes Marching On).

And then we went upstairs to the hearing rooms. Besides the nuclear activists (such as myself), many pro-nuclear people from Central New York came to this meeting in Albany. Tim Knauer's article on Syracuse.com has a picture of part of the scene at the rally: "Dozens of CNY residents flood Albany meeting on nuclear subsidies."[189]

I had the good luck to get a seat in the Public Service Commissioner meeting room itself: the department had to open three "overflow" rooms with video feeds because of the large crowd. I heard the historical decision to support all kinds of clean energy—renewable *and* nuclear.

I love the picture of us celebrating after the rally (figure 14). We are gathered outside the building in which the meeting was held. Michael Shellenberger has opened a bottle of non-alcoholic bubbly. He shook it first, so it is spraying all over. Eric Meyer's face is partially hidden by the bubbles. Rod Adams is at the right, with his fist in the air in joy. Sarah Woolf is holding her "Mothers for Nuclear" sign at the left. I'm there with my pro-nuclear T-shirt from my own rallies.

People came to Albany to support nuclear energy. People came from the New York upstate plants; they came from all around New York and New England; they came from Virginia.

The rally is a wonderful memory for me. And the upstate nuclear plants are still running.

Figure 14: August 1, 2016 rally in New York State, celebrating the ZEC. Picture courtesy of Steve Whiting.

New York was the first state to explicitly support zero-emission sources, including nuclear. More than half of the low-emissions power in the United States is nuclear power.

The New York vote to support nuclear power was the equivalent of a Renewable Emission Credit for nuclear, but at much less cost per kWh than most RECs. This did not happen because of some sort of a sudden realization by the New York Public Utilities Commission that they had been wrong to pick and choose low-emissions sources to support. As we have seen with Massachusetts having six types of RECs, states pick and choose technologies all the time. With initiative, patience, and organizational skills, nuclear supporters helped to achieve good results.

ZECs and New York energy

ALL POLITICS IS LOCAL. The New York Public Utilities Commission did not get up one morning and realize that there were good technical reasons to support nuclear power. What actually happened is that three nuclear plants in upstate New York were in danger of closing, due to competition from natural gas and from renewables able to sell power at less than zero cents per kWh. In other words, the renewables were able to pay the grid to take their power. This business model (I'll pay you to take my product) was possible due only to the various RECs and PTC subsidies that renewables obtain.

If those three nuclear plants had closed, many jobs would have been lost in upstate New York, and there would be a serious hole in the upstate power supply. Nuclear workers held rallies at the state capital, had letter-writing campaigns, and were supported by pro-nuclear people from many areas of the country. I was one of those people. I was proud to be one of the nuclear supporters, a person with something to celebrate.

The New York Department of Public Service passed the ZEC ruling. Non-nuclear power generators challenged the ruling in court. Ultimately, the U.S. Court of Appeals for the Second Circuit ruled that New York's Clean Energy rules were constitutional and within the legitimate purview of a state.[190]

One of the most amazing things about this ruling, for me, was that the Environmental Defense Fund (EDF), usually no friend of nuclear, supported the ZECs. The EDF even filed an amicus brief with the court.[191] I don't think that EDF suddenly understood the important role of nuclear. I think they realized that, if states don't have the right to set up clean-energy rules of whatever type, the states' renewable portfolio standards and the states' manifold types of renewable RECs could also be challenged in court. I think that EDF didn't want renewable-support rulings to fall in court, and nothing would hold those rulings up if the ZEC ruling fell to a lawsuit.

Meanwhile, however, Governor Cuomo and the New York legislators continued to do their best to harass Indian Point (the big downstate plant) into closing. As I write this, they have partially succeeded: Unit 2 of Indian point was shut down on April 30 2020.[192] In my opinion, the difference between the upstate and downstate plants is about politics. Governor Cuomo wanted to play to his anti-nuclear New York City base by shutting Indian Point, but he did not want to alienate the upstate New York pro-nuclear counties by shutting down their plants. In practical terms, the ZEC rules accomplished this by setting the ZECs so that Indian Point would not be eligible for the payments. Indian Point sells its power into a higher-priced market, and the ZECs have a cut-off price. (If you are already getting this price per kWh, you don't get ZEC support.) Energy policies frequently follow political agendas. It is easy to write the laws so that politics wins.

I don't want to follow the internal politics of New York State much further here. At this point, New York State, Illinois, and New Jersey have some version of ZECs to support at least some of their nuclear plants. Other states, such as New Mexico and Connecticut, have nuclear set-asides or clean-energy policies that include nuclear energy.[193] I hope this type of policy will spread. As I am writing this, both Pennsylvania and Ohio are considering similar bills promoting ZECs.

This book is about grid policies. If you have grid policies that supply subsidies to low-emissions renewables, you can also have grid policies that supply subsidies to low-emissions nuclear.

My hope that ZECs are adopted in more states is also strongly about the environment and the grid. Let's look at Pennsylvania and Ohio first, since they are the states currently considering ZECs.

In 2017, Ohio made 57% of its electricity from coal and 15% from nuclear. Pennsylvania made 39% of its electricity from nuclear and 22% from coal. This information comes from a *New York Times* article based on U.S. government information.[194] As an environmentalist, I would rather see coal plants shut down before nuclear plants are shut.

Also, as I have described before, a grid completely dependent on just-in-time gas is a very vulnerable grid. Low-emissions nuclear plants add stability. If renewable plants can get heavy subsidies for their power, low-emissions nuclear plants can help the environment and the grid, with fairly small levels of subsidy.

What do I mean by "fairly small levels of subsidy"? At the time that the New York State ZEC was passed, subsidies (federal and New York) for renewables were 4.6 cents per kWh, and the ZEC subsidies for nuclear energy were 1.7 cents per kWh.[195] ZEC payments are part of the New York Clean Energy Standard. When announcing the Clean Energy Standard, Governor Cuomo estimated that it would cost less than $2 a month for the average residential customer.[196]

How does $2 a month compare with other costs for subsidies? This is hard to calculate, unless you are an expert in energy tariffs. However, in chapter 31, "RECs Are Us," I referenced James Bride's calculations of the costs of compliance with renewable requirements in New England. The costs varied by state, from 0.2 to 1.6 cents per kWh. For comparison, I'll do a quick calculation.

First, we'll assume that the average cost of renewable compliance is 0.9 cents per kWh. This is the average of the lowest cost of compliance (0.2 cents per kWh, Maine) and the highest cost of compliance (1.6 cents per kWh, Massachusetts) from the Bride study.[197] Next, I took the average U.S. household use of 914 kWh per month from the U.S. Energy Information Administration.[198] Multiplying the compliance-cost average (0.9 cents per kWh) by the average monthly household usage (914 kWh per month per household), the average cost for renewable compliance was $8.23 per month per household.

These calculations were easy, but the numbers are not universal. Many states with heavy air-conditioning use will undoubtedly have higher kWh usage per household. States outside of New England may have lower "costs of compliance" with their less-extensive renewable mandates. Still, about $8/month is a start at estimating the costs of renewable compliance and a beginning for understanding what I meant by "fairly small levels of subsidy" for the New York State ZEC, at $2 a month.

Subsidies on the grid can become very complicated and convoluted. Before we look at some really complicated subsidies for renewables and gas-fired plants, let's look at what nuclear ZECs are buying for ratepayers. Let's look at emissions.

A note about emissions

How low are emissions from nuclear energy? According to the Intergovernmental Panel on Climate Change report of 2014, the life-cycle greenhouse-gas contribution of nuclear is 12g of CO_2/kWh, about the same as offshore wind and lower than utility-scale solar PV at 48g/kWh.[199] (The IPCC report is a meta-study summarizing several estimations; I have chosen the medians for my comparisons.)

We look at life-cycle emissions because a nuclear plant or solar panel does not emit appreciable carbon dioxide while it is operating.

Nuclear and solar are not combustion processes. However, building the plant or making the solar panel requires some level of fossil fuels, so the proper emission comparison is life-cycle emissions.

The combined-cycle natural-gas estimate in the same IPCC report is 490 g/kWh, about forty times higher than the nuclear estimate. Making electricity with natural gas is a combustion process, which produces one molecule of carbon dioxide for every molecule of methane burned.

To see the effect of shutting down a nuclear plant, we could look at what happened when Vermont Yankee shut down. Gas-fired energy replaced Vermont Yankee.

Mike Twomey of Entergy calculated the replacement using ISO-NE data. When Vermont Yankee shut down, its power was replaced pretty much kWh for kWh by natural gas.[200] In 2014, with Vermont Yankee running, nuclear energy made 36.4 million MWh on the New England grid, and natural gas made 46.2 million MWh. In 2015, after Vermont Yankee shut down, nuclear made about 4.9 million MWh less energy, and natural gas made 5.7 million MWh more energy. (Natural gas also replaced a coal plant, with coal's contribution going down by 1.2 million MWh.) Meanwhile, solar's contribution went up by 0.1 million MWh, and wind's contribution went up by 0.3 million MWh.

Yes, it was gas that replaced Vermont Yankee.

To find out how much CO_2 closing Vermont Yankee added to the atmosphere, let's look at those life-cycle figures again. Natural gas made 4.9 million MWh more than it had made the year before. This turns into 2,400,000 metric tons of extra carbon dioxide added to the atmosphere in one year, due to the nuclear plant closing. I have put the calculation in an endnote.[201]

The amount of carbon dioxide added is undoubtedly higher than that. For traceability of my data, I am using the IPCC medium level of CO_2 emitted by a combined-cycle plant.

Not all the gas plants on the grid are combined cycle; most of them are actually single cycle, which are much less efficient in turning a cubic foot of gas into a kWh of electricity. A typical single-cycle turbine has a thermal efficiency of 35 to 40%,[202] while a combined-cycle gas turbine has thermal efficiencies between 55 and 59%.[203] Using the higher efficiency numbers for both types of turbines, we can compare the single-cycle and combined-cycle turbines. To produce the same amount of power, the single-cycle turbine would have to burn 1.48 times as much gas as the combined-cycle turbine would require.

Both types of gas turbines are on the grid. However, the IPCC numbers are for combined-cycle turbines. Using these numbers meant that turbine efficiency was overestimated and that carbon-dioxide emissions were underestimated. Because there are many single-cycle gas turbines in use, 2,400,000 metric tons is an underestimate of the amount of carbon dioxide that went into the air per year because the nuclear plant closed.

CHAPTER 34

RENEWABLES AND AUCTIONS

MOPR AND CASPR ARE NOT the names of two adorable basset hounds, though a friend suggested that these would be great names for sweet dogs. MOPR is Minimum Offer Price Rule, and CASPR is Competitive Auctions with Sponsored Policy Resources. These programs describe two ways that ISO-NE manipulates the capacity auctions. MOPR rules are also a major issue in other parts of the country.

Energy, capacity, states, and so on

IN THE LAST FEW chapters, I described some state policies encouraging renewables, some new state policies encouraging nuclear, and a little bit about how these policies intersect with RTO policies, which are required to be fuel neutral. We've seen all the trouble the RTOs can have with fuel neutrality when it comes to keeping the lights on in winter, but the RTOs can't just say, "This set of plants ... store some oil on-site."

In the Second Jump Ball chapter, we ended up with the ISO-NE Pay for Performance, a program that transfers funds from capacity payments made to one plant to capacity payments made to a second plant. If a plant doesn't come online when the grid is stressed, it has to pay part of its capacity payment to a plant that did come online.

Many groups objected to Pay for Performance because plants would be vulnerable to losing part of their capacity payments, and, therefore, they might alter their bid to get more payment for capacity in the first place. This business practice is called a "risk premium" and somehow upsets RTOs, who want a market but not with plants charging more if they take more risk. (Hint ... it isn't a "market" if you can't charge more if you take more risk. As I have noted before, what the RTOS call a "market" has little resemblance to a market.) At any rate, with Pay for Performance, the capacity payments are likely to fluctuate.

There was pushback. There were competing interests.

The next program that ISO-NE proposed was also about capacity payments, but in a different context. This program was about keeping capacity prices high and not letting subsidized resources undercut the pricing in the capacity markets. The new program is called MOPR (Minimum Offer Price Rule). MOPR basically says that subsidized resources cannot consider their subsidies in their capacity bids. They have to bid into the market at a minimum price.

Guess what. There was pushback on MOPR, so CASPR (Competitive Auction for Sponsored Policy Resources) was invented. And so it goes, as more and more rules are layered onto the supposed "market."

Basically, renewables get "outside of market funding" through production tax credits, selling RECs, and so forth. In several states, nuclear plants also get such funding in the form of Zero Emission Credits, or ZECs. Whether it is renewable energy selling a REC (renewable energy certificate) or nuclear energy receiving a ZEC payment, these payments are kind of the same thing. Depending on state rules, some types of power plants get extra payments, which usually have nothing to do with the role of the plant on the grid. These payments are called "out-of-market" payments, and they are usually set by state policies. Most of the out-of-market payments are meant to encourage low-emission plants.

The state-policy payments are becoming more and more important. Sometimes, only renewable plants can make money, because of their out-of-market payments. In terms of the RTO rules, these subsidized generators are often called "Sponsored Policy Resources."

As usual, once you begin weaving a web of regulations, you need more regulations to fix things. For example, instituting Pay for Performance meant that power plants are no longer guaranteed their capacity payments. The capacity payments were put in place to make sure that power plants would be kept in decent shape to run, and, therefore, they could run when called upon. Pay for Performance took away the capacity-payment guarantee. Then it was noticed that the plants would charge more for their kWh or raise their bids for capacity payments, and add a "risk penalty," if they didn't have a guarantee. This would probably raise the ultimate price to the consumer. Pay for Performance is described in chapter 18, "Second Jump Ball and Pay for Performance."

No doubt some complex "fix" for this new problem is in the works. The RTOs are probably working on a new rule, a rule which will guarantee plants sufficient capacity payments to keep up their maintenance, and simultaneously guarantee that the capacity-payment penalties are sizable enough so that they will arrange to run when called on. There will be guarantees but also penalties. These will be figured out "just so" for all the plants. Personally, I am not holding my breath for this to happen. The tweaks aren't working, and the just-so stories are failing.

In 2016, Entergy and First Energy announced that they are leaving the RTO areas because their plants could not make enough money to keep running. As I wrote in a 2016 blog post,[204] First Energy was following the Entergy pattern of selling or closing plants in the RTO areas and operating power plants only in vertically integrated areas. As a matter of fact, when Entergy exited the RTO areas, they even sold a gas-fired plant. These two companies no longer want to own plants that compete in the RTO areas, partially because the plants compete with renewables that receive out-of-market payments for their energy. In an RTO area, selling kWh is a losing game.

Selling capacity, however, can be a winning game. A recent paper by Jacob Mays et al. describes how capacity payments affect the markets.[205] (The authors are at Northwestern University and FERC.) The title of the paper is "Asymmetric Risk and Fuel Neutrality in Capacity Markets." It describes how the introduction of capacity markets in RTO areas "has an asymmetric effect on the risk profile of different generation technologies, tilting the resource mix toward those with lower fixed costs and higher operating costs." In other words, according to the careful analysis in this paper, a capacity market decreases the economic risk of operating a plant with low capital costs but relatively high fuel costs (such as a single-cycle gas turbine) but increases the economic risk of operating a plant with high capital costs but low fuel costs (such as a nuclear plant).

An example calculation from this paper shows that a plant in the PJM RTO area that uses inexpensive fuel can expect to earn 17% of its operating profit from the capacity market, while a plant that uses expensive fuel can expect to earn 90% of its profit from the capacity market. This conclusion is closely in line with the graph in figure 8 (chapter 15, "Selling kWh Is a Losing Game"). That graph shows the percentage of revenue that different types of plants get from different sources. High-priced "peaker" plants (gas-fired) get most of their revenue from the capacity market and will probably fight to the death to defend their capacity payments.

Moving from energy to capacity

SUBSIDIZED RENEWABLES CAN distort the energy markets. Nuclear, coal, and combined-cycle gas plants were affected by renewables bidding low on the energy markets (due to renewables' out-of-market funding). However, gas was sitting pretty, no matter how low the energy-market prices dipped, because gas received most of its money from capacity payments.

But wait: Renewables can also bid into the capacity market.

However, there are often penalties for receiving capacity payments but not going online when called upon to supply electricity. Since wind turbines are well aware that they can't order the wind to blow, a wind turbine will generally not bid in to the capacity market at anything like its nameplate capacity. A wind turbine does not want to be charged a large penalty for not going online when called by the RTO. Therefore, wind turbines often bid into the capacity market at "capacity value," less than their nameplate capacity.

This factor-value stuff takes some explaining. The *capacity factor* is the fraction of time the wind turbine can be online. It counts the time that the wind is blowing in a range of speeds which can produce electricity, whether or not the wind turbine's energy is needed at that time. In contrast, the capacity *value* is the fraction of time the wind turbine can produce electricity that can actually be used on the grid.

By bidding in at a low "capacity value," a wind turbine will receive some capacity payments, but they won't get hit too badly with fines if the wind doesn't blow. That is the wind turbine's way of looking at the capacity value, perhaps. From the grid's point of view, capacity value estimates how valuable the wind turbine is to the grid. For example, if a traditional power plant bids into a capacity auction, it generally bids at its nameplate capacity, or close to it. If a wind turbine bids in at 10% of its nameplate capacity, it is pretty much admitting that it is only 1/10 as valuable to the grid as a standard dispatchable power plant.

How to calculate "capacity value" for wind is a thriving academic study. For example, PJM (the RTO that includes Pennsylvania, New Jersey, and Maryland, among other areas) calculates the "effective load-carrying capability" of wind. This is basically an estimate of wind's capacity value.

For their calculation, PJM looked at hourly load shapes as well as the wind-availability shapes. According to a PJM report in September 2018,[206] the mean effective load-carrying capability for wind was 11.5% of nameplate capacity. That number is a type of estimation of wind's capacity *value*. In contrast, wind capacity *factors* range between 22% and 45% (nationwide figures from EIA).[207]

As described earlier in this book, wind turbines receive significant out-of-market funding. That means they can underbid conventional plants on the energy markets. It also means that wind can underbid conventional plants in the capacity markets. This didn't matter when there were very few wind turbines, but as more and more wind turbines are added to the grid, wind-turbine bids can affect the capacity market and begin to lower capacity payments for "peaker" gas plants.

MOPR to the Rescue

WHEN WIND BEGAN affecting capacity payments for gas-fired plants, clearly, something had to be done! (I am using a passive voice here. I think the fact that capacity payments were at risk—and, therefore, gas plants were at risk—was an interesting "trigger" for ISO-NE and FERC market intervention. But I am not saying who noticed or pulled the trigger.)

FERC allowed ISO-NE to implement a MOPR, a Minimum Offer Price Rule, for capacity bidding. There are variations of this rule at other RTOs, also. In a MOPR, according to the FERC ruling:[208]

> ISO-NE utilizes a minimum offer price rule, or MOPR, that requires new capacity resources to offer their capacity at prices that are at or above a price floor set for each type of resource (referred to as the Offer Review Trigger Price). The MOPR does not allow resources receiving out-of-market revenue to reflect that support in their offer prices, unless the support is widely available to other market participants.

Since renewable power plants (such as wind turbines) get REC payments and other out-of-market payments, they do not need to get much money from the capacity markets. In the energy markets, a wind turbine can bid in very low (zero or negative) and still make enough money from out-of-market payments. The same is true in the capacity markets: the wind turbines can bid in very low.

Just as zero bids in the energy markets will lower the clearing price for MWh, low bids in the capacity markets will lower the clearing price for power. (Usually capacity is sold as power availability for a month at a time. This stuff is impossibly complex.)

MOPR provisions are controversial: According to an April 2018 *Utility Dive* article, FERC Commissioner Neil Chatterjee does not support the part of the FERC ruling that establishes MOPRs as the "standard solution" to deal with state power subsidies.[209] Chatterjee said he voted for the ruling because it was part of a two-part capacity auction. (The other part was CASPR.) But we are getting ahead of ourselves by discussing CASPR. Let's stick to the reasons for MOPR for a few more sentences.

Why would low capacity prices (due to subsidies) be a problem that ISO-NE is concerned about, while low prices in the energy markets (due to subsidies) are merely business as usual? Low energy prices hurt some kinds of plants (traditional, baseload-type plants), and low capacity prices hurt other kinds of plants (mostly gas-fired load-following or peaker plants). Therefore, I could make some cynical comments here. Instead, I will merely think my comments. I will also think a bit about how ISO-NE always has to look over its shoulder at the NEPOOL Participants Committee. That committee seems to be dominated by gas-fired plants, as far as I can tell.

Okay, enough of my follow-the-money reasoning here. Now I will go to the more usual explanation for why ISO-NE is concerned with low *capacity* prices due to subsidies, while it is blissfully unconcerned with low *energy* prices due to subsidies.

ISO-NE fears that low prices in the capacity markets will mean that new power plants won't be built. Capacity payments are supposed to encourage new builds by providing revenues that are close to the Cost of New Entry (CONE)—that is, the cost of a new power plant. When a company decides to build a plant, it will bid into the capacity market. If its bid is accepted (its bid isn't too high), it can bank on capacity payments when it comes online. As mentioned before, capacity payments are the major way that many gas-fired plants are funded, so a guarantee of such payments can encourage bank loans for building the power plant. The capacity payments also encourage profitability once the plant has been built.

However, if the market clearing price for capacity payments is too low, perhaps this new plant won't be promised enough funding, and it won't be built. Perhaps no new gas-fired plants will be built. To fix this, ISO-NE proposed the MOPR rule to FERC. The idea of MOPR is to keep capacity payments high enough to encourage CONE (new power plants).

Versions of the MOPR rule have also been suggested in some other ISOs, but not all. PJM has proposed an MOPR rule to FERC, but, as of this writing, FERC has not yet ruled on it. However, FERC postponed the August capacity auction at PJM. At first, FERC allowed PJM to hold its capacity auction under the old rules.[210] More recently, FERC postponed the auction: Commissioner Glick was concerned that the auction would not even be legal.[211]

Meanwhile, in ISO-NE, capacity prices have gone higher, even as energy prices fall. As shown in figure 15, energy payments in 2008 were $12.1 billion while capacity payments were $1.5 billion. In contrast, energy payments in 2018 were $6.0 billion while capacity payments were $3.6 billion. In other words, in 2008, capacity payments were about 12% as much as energy payments while, in 2018, they were 60% as much.[212]

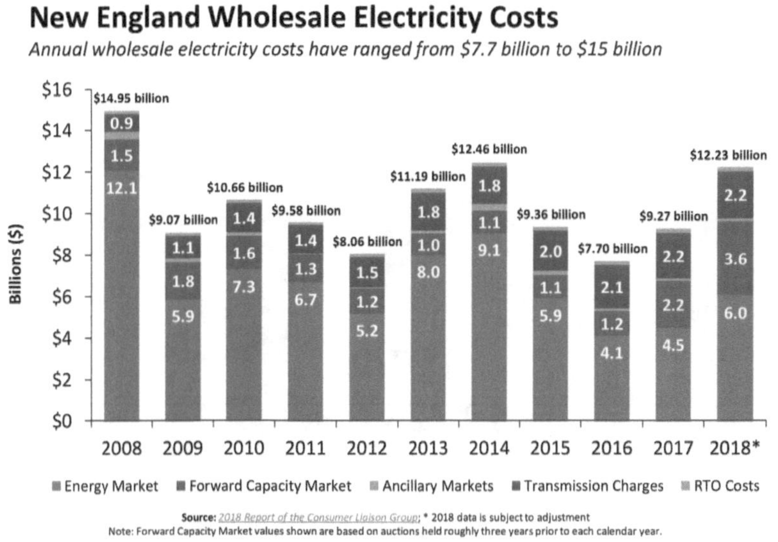

Figure 15: New England wholesale electricity costs. (ISO-NE)

In this situation, plants that rely on capacity payments (like gas plants) are the winners, and plants that rely on selling power (baseload plants) are the clear losers.

CONE is set up to encourage plants that don't have high capital costs and for whom capacity payments are the major part of their payments. CONE is not expected to be high enough to support the capital costs of a nuclear or coal plant. However, CONE can support a natural-gas plant. Looking at figure 16 from the June 2019 presentation by Ann George of ISO-NE,[213] we can see the new plants that are proposed for ISO-NE over the next three years. The queue consists of 14,000 MW of wind and solar, 3,000 MW of natural gas, and 2,000 MW of batteries. In short, the newly proposed plants are running about a three-to-one ratio of intermittent renewables (14,000 MW) to firmly dispatchable power.

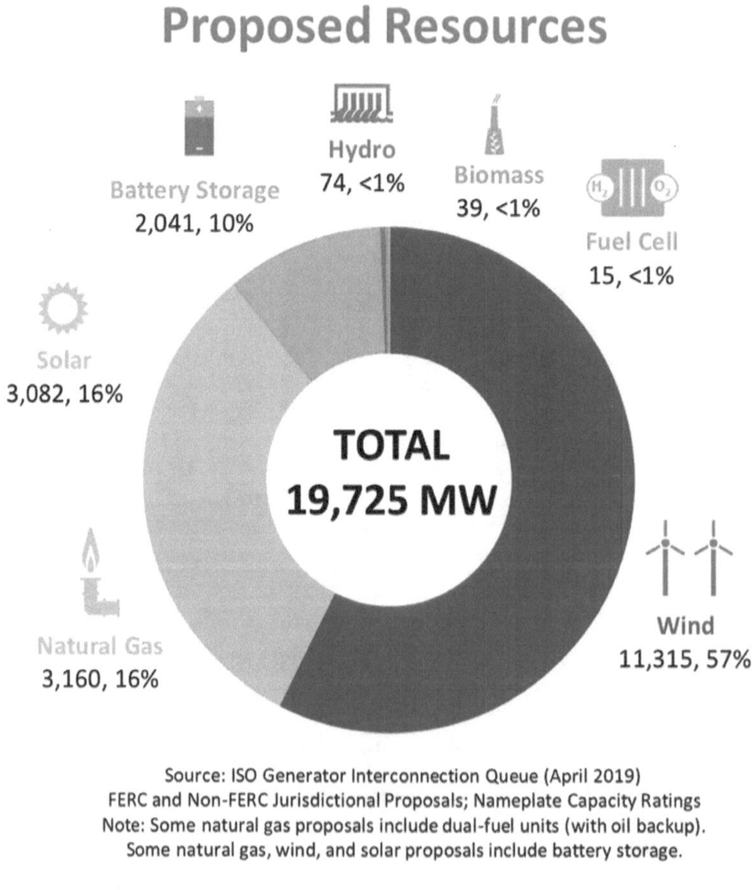

Figure 16: The ISO-NE queue: 11,000 MW wind and 3,000 MW natural gas (ISO-NE)

MOPR, CASPR, CONE

WHEN *ISO-NE* SAW SO many power plants with out-of-market funding (like wind turbines) in the queue, they were concerned that these plants could bid very low into the capacity auctions and, thereby, suppress capacity prices. These low prices could affect the capacity market's ability to retain and attract resources that do not have out-of-market funding. ISO-NE asked to change their tariff rules to encourage reasonable price signals. In March 2018, FERC ruled in ISO-NE's favor.[214] The quotes below are from pages 2 through 5 of the FERC ruling.

Describing their existing rules, FERC writes that "ISO-NE utilizes a minimum offer price rule, or MOPR, that requires new capacity resources to offer their capacity at prices that are at or above a price floor set for each type of resource (referred to as the Offer Review Trigger Price)." These Forward Capacity Market MOPR rules do "not allow resources receiving out-of-market revenue to reflect that support in their offer prices."

However, those same rules "permit a limited exemption from the MOPR for certain renewable resources (the Renewable Technology Resource (RTR) exemption)." This exemption allows 200 MW of renewable resources to bid into the Forward Capacity Auction at very low prices, reflecting the out-of-market subsidies they receive. When 200 MW of renewables do not bid into one year's auction, the allowed RTR capacity bidding carries over to the next year. For example, if there were no RTR bids one year, there could be 400 MW worth of RTR bids the next year. The maximum RTR bidding in a given year is capped at 600 MW.

But RTR won't do enough. New England states have developed very aggressive renewable-energy goals. As described in the FERC ruling, ISO-NE had noted that "the 2016 Massachusetts Energy Diversity

Act ... requires clean energy procurements in the range of 2,800 MW (nameplate)." Clearly, the new renewable goals are far higher than the RTR exemption could accommodate. In January 2018, ISO-NE filed to change the rules in order to be able to accept more resources that receive out-of-market funding, while still maintaining "competitive capacity pricing."

As FERC described the situation in its ruling:

> According to ISO-NE, these out-of-market actions [by the states] could result in price suppression and thus negatively impact the market's ability to retain and justly compensate needed existing resources and to attract new, competitively compensated resources. Another concern, ISO-NE states, is that the MOPR may cause consumers to 'pay twice' for the same capacity—i.e., pay once for capacity procured in the FCM to serve their demand and pay a second time for the additional capacity obtained through out-of-market contracts with state-supported resources.

MOPR would help the gas-fired plants compete with renewables, and MOPR would increase the stability of the grid. But what about those renewables? They will be out-competed by the gas plants, perhaps. Not good for renewables. To keep things sort of even (if this is even possible), ISO-NE proposed CASPR: Competitive Auctions with Sponsored Policy Resources.

CASPR turns the Forward Capacity Auction into a two-part auction. In the first auction, plants that don't get significant subsidies will compete under MOPR rules. In the second auction, plants that do get significant subsidies will be allowed to consider those subsidies in their bid to the second auction, the CASPR auction. Therefore, subsidized plans will compete at very low bid prices in the CASPR auction.

When the CASPR auction is concluded, there can be a sale of capacity obligations between different types of plants.

The plants that competed only in the MOPR auction, the plants that do not get significant subsidies and, therefore, get MOPR (higher) capacity payments, can look at the plants that competed in the CASPR auction. Then the MOPR auction plants can sell their capacity obligations to subsidized plants at the low price that the subsidized plants received in the CASPR auction. Once the MOPR plant has sold its capacity obligation, the MOPR auction plant can retire.

After the CASPR auction, plants that plan to retire will pocket the difference between what they received for a certain number of MW in the MOPR auction and the amount they have to pay for the same amount of MW to a subsidized resource who bid into the CASPR auction. The difference is a sort of retirement subsidy.

This two-step MOPR-CASPR auction makes it easier for a non-subsidized existing plant to retire: it pays CASPR (subsidized) plants to take over its capacity obligations. Meanwhile, the non-subsidized existing plant pockets the MOPR money, which is by definition higher than the CASPR payment. The two-step auction means that renewables will not lower the overall capacity price on the grid but will compete only in their own auction. Renewables will not lower the capacity price on the grid. ISO-NE has protected CONE. Plus, CASPR makes it easier for an existing plant to retire.

The *theory* is that CASPR makes it harder to overbuild capacity and, therefore, saves money. The *reality* is that CASPR offers a way for an existing plant to shut down and get a small bonus for doing so. Their bonus is the difference between the money the plant had received in a MOPR capacity obligation and the money it will pay to a renewable resource who has outside funding and was allowed to bid low in the low-priced CASPR secondary auction.

CASPR holds the capacity prices high. It protects the major funding stream of the gas-fired plants.

It's funny how things always seem to work out this way.

CASPR and FERC

CASPR WAS APPROVED BY FERC in March 2018. At that time, Robert Powelson was a FERC Commissioner, appointed by President Trump to fill one of the "Republican" slots on the Commission. (The Commission usually has three Commissioners from the president's party and two from the opposition party.) In March 2018, Powelson dissented from the majority opinion on MOPR and CASPR. In August of 2018, Powelson resigned to be president of the National Association of Water Companies.

Powell's resignation came after less than a year of service as FERC Commissioner. Some commentators did not believe Powelson's stated reason for his resignation, which was that the Water Association position was a "tremendous opportunity." The *Washington Examiner* headlined that "Republican FERC commissioner Robert Powelson to resign amid fight over coal bailout."

In terms of his dissent about CASPR, however, I think the crucial point is that Powelson is a "fierce defender of competitive markets" (as described by the *Washington Examiner*).[215] As such, he dissented from the ruling that allowed CASPR. His dissent was important and cogent, and I quote from it extensively below:[216]

> In recent years, however, certain restructured states in New England have again taken an active role in resource planning by attempting to procure select types of generation resources in their state. These efforts began with the enactment of renewable/alternative energy portfolio standards and, more

> recently, have involved other forms of out-of-market support for select resources.
>
> There is no question that states are entitled to procure any resources they prefer. It is important to note, however, that no New England state has signaled a desire to change current responsibilities for resource adequacy. Grid reliability and resource adequacy remain within the purview of the regional grid operator, and it is the Commission's responsibility to ensure that this objective is accomplished at just and reasonable rates.
>
> The CASPR proposal appears attractive because it ... presents a solution to this complicated situation. However, the major flaw in the proposal ... is that the "competitively based" market clearing price will not provide a meaningful signal to the marketplace.
>
> I am further concerned about the signals that today's decision sends to New England stakeholders. Instead of incentivizing developers to compete for market revenues, the message the Commission is sending to market participants is that the best way to ensure the future viability of a particular resource is to seek state support. This is not a prudent policy choice.
>
> Critically, CASPR will not be a final resolution to the problem.

I agree with Powelson. CASPR is one more band-aid on an unfixable situation. State policies and out-of-market payments distort whatever power and capacity markets exist in the RTOs. Instead, the states happily make policies, favoring their favorite virtuous technologies.

The states force the RTOs to come up with band-aids to keep the grid running. People hope that the band-aids will lead to grid reliability and low-cost electricity.

However, nobody (not the states, not ISO, not the plant owners) is responsible for grid reliability or expense. The buck never stops. When things go wrong, the entities (states, plant owners, ISOs) don't even have to bother passing the buck and shifting the blame. They never had the responsibility, so they don't have the blame. The grid may be in trouble, but they aren't.

The grid needs an equivalent of Harry Truman. As president, he had a sign on his desk: "The Buck Stops Here."

All the RTOs seem to be heading down the "multiple capacity auction" route. A recent article by Iulia Gheorghiu in *Utility Dive* provides a good compare-and-contrast of various RTO plans.[217]

The descriptions of these auctions describe the "carve-outs" and the search for "granularity" in capacity auctions. For example, FERC recently turned down a PJM proposal for a two-step auction with a "resource carve-out plan." MISO, the Midcontinent Independent System Operator, is the RTO for much of the Midwest and some of the South. It has a capacity market in the southern part of its territory but only a "voluntary residual" capacity market in its northern areas. New York ISO's capacity market is seasonal. In all these cases, the RTOs are basically reacting to states' renewable mandates.

Unfortunately, Powelson had it right. No matter how careful a fix is proposed for the capacity auction, these band-aids send the wrong message. As quoted above:

> Instead of incentivizing developers to compete for market revenues, the message the Commission is sending to market participants is that the best way to ensure the future viability

> of a particular resource is to seek state support. This is not a prudent policy choice.

In other words, in the RTO auction areas, the rules for renewables will force all power plants into seeking state support in order to remain viable. The ones that don't get support quickly enough will be most likely to close. Retail costs for electricity will rise, but grid reliability will fall.

This isn't how a market is supposed to work. But the RTO areas are more of a set of band-aids than a market.

THE RTO AND THE CUSTOMER

CHAPTER 35

THE LARGEST MACHINE ON EARTH

The grid as machine

THE AMERICAN ELECTRIC GRID has been called "the largest machine on earth." Even looking at a single RTO area or a large area served by a balancing authority (such as Bonneville Power), we can be looking at hundreds of square miles of territory, tens of generators, hundreds of substations, and thousands of miles of wire. A grid is a big machine.

A grid is made up of many machines (generators, switches, wires), but they all have to work together. Your car is a smaller machine, and your car wheels have to work together so that the car can go around a corner. Similarly, the parts of the grid have to work together. The grid parts work together so that electric load and generation are always closely matched, the frequency of the grid remains within strict limits, the VARS are okay, N-1 reliability is maintained, and so forth. The electric grid is the pre-eminent engineering achievement of the 20th century.

But the grid is not just "so twentieth century, and old hat." At the beginning of the twentieth century, people kept food cold with ice boxes. Ice was cut from northern lakes and stored in sawdust for the summer. At the beginning of the twenty-first century, it would have been a very poor home in America that didn't have a refrigerator with a freezer included. No, the grid is not "old hat."

When your car was designed, the first step was the specifications. Then came designing a machine to meet those requirements, procuring parts and testing prototypes, and then, finally, setting up a manufacturing line to build your car. Unfortunately, with the grid, several of these steps have been skipped. Well, no, they aren't completely skipped. However, the planning stage is no longer considered very important.

Utilities used to have to file Integrated Resource Plans (IRPs) with their state PUC. According to a 2015 blog post by Coley Girouard on the industry website Advanced Energy Perspectives, 33 states still require utilities to file Integrated Resource Plans with their state PUCs.[218]

An IRP basically consists of a load forecast and plans for how the load will be met. The plans include power plants built, retired, and upgraded; transmission lines built, retired, and upgraded; expected demand-side management possibilities; and so on. In other words, for the next twenty years, how are you (the utility) going to meet the demand for electricity, what will have to be built or upgraded, and how much will it cost?

These plans were generally publicly available (by legal requirement). I will also say that, by its nature, an IRP can be understood by a thoughtful layperson. Yes, an IRP could be long and full of legalese, but in general, the plan is: we expect this amount of electric usage, and this is what we plan to do about it. The plan is essentially understandable by the citizen.

In contrast, it takes a long time to explain how ISO-NE is attempting to protect the capacity auctions providing enough money for CONEs by a complex two-step forward-capacity-auction process, the MOPR-CASPR two-step. The choices involved in RTO forward-capacity auctions are not easily understandable nor are Pay for Performance or the Beat the Peak days. RTO decisions cannot be understood unless a person has studied the thick and ever-growing tariffs of the RTOs. In other words, unlike an IRP, the RTO choices are opaque to the citizen, even when reporters are allowed to be at their meetings.

The rise and fall of the IRP

IRPs WERE VERY POPULAR in the '80s. With the formation of RTOs and their auctions, people thought that, perhaps, the IRPs would go away, and the market would take over. As the Girouard blog post says, however, the energy crisis in California (around 2000) "gave many regulators pause and led them to believe that some level of integrated planning was still a necessity."

The IRP did not go away, but its scope has been deeply diminished in the RTO areas. The forward-capacity auction and its CONE are supposed to determine which new plants are built. The states have comparatively little to say about this. However, states can and do make rules such as "no new nuclear plants" or "dual-fueled gas plants can burn oil for only a certain number of hours each year." States even have IRPs, though in the RTO areas, the RTO processes tend to override most of the states' power.

As we have seen earlier in the "Energy, Capacity, States, and So On" section of chapter 34, RTO rules protect gas-fired plants financially. Even MOPR and CASPR protect gas-fired plants: the Cost of New Entry, the CONE, is hopefully enough to build a gas-fired plant.

Meanwhile, many state legislatures, both in RTO areas and elsewhere (but mostly in RTO areas) exercise their power by ordering their in-state utilities to buy a certain percentage of power from renewables. In an RTO area, there seems to be the idea that the RTO is taking care of the grid as a whole. State legislatures don't feel they have to worry about reliability: the RTO does that. (Actually, the RTO has very limited powers to assure reliability. Remember the three new auctions proposed by ISO-NE?) Even costs will be determined by the "market."

In many areas, the RTO is also theoretically responsible for system planning. However, since the RTO has to be fuel neutral, and the auctions are supposed to be the "market" that the RTO is operating, the RTO actually has very little power to implement plans. Earlier in this book, we saw the New England RTO struggle to design a method to keep the lights on in winter. ISO-NE tried to arrange for dual-fuel plants to have oil on hand. Describing these battles took several chapters, and that fight was only about stockpiling oil for emergencies. It would be close to impossible for an RTO to plan and implement a resilient mix of power plants and fuels. A single-state Public Utility Board could insist on a strong and resilient IRP plan: an RTO can't.

But the RTO auctions do provide cover for state mandates. From the point of view of a state legislator, why not feel good and vote for a law that says the state electricity will be X percent renewables by year Y? It doesn't seem to hurt anyone, and it may get you some votes in some areas. Voting for a "high percentage of renewables" is voting for the modern version of "motherhood and apple pie." Meanwhile, the RTO is in charge of the grid. If someone begins to say that so many renewables may cause problems, the state legislator can say that a single state's plan is no big deal for the grid as a whole. The RTO will handle it.

In the days when states used integrated resource planning, utilities had to justify their expenditures to their PUCs. They had to show that they were putting together a proper mix of power sources, so people in the state would not be blindsided by a sudden rise in cost of one fuel or expensive new environmental laws affecting another fuel. The lack of such state oversight is probably one of the reasons RTO areas tend to have higher consumer prices than non-RTO areas. (I cover the comparison studies in chapter 13, "The Death of the Market Hope.")

"Motherhood and apple pie" does not convince a PUC board that a utility is spending the consumer's money wisely.

The costs show up fastest in a small area. As I have observed, the rural electric cooperatives in Vermont are the utilities that are most concerned with price rises due to renewable subsidies. In Vermont, one of the early strong proponents of net metering, Tony Klein, is now concerned that net metering costs may drive some of the smaller cooperatives to the brink of bankruptcy.[219]

In a rural cooperative, the generation owners and the power customers are the same people, and the area is small. An RTO area is much bigger, and the generation owners and the customers are not the same people. But the extra costs of renewables exert the same market forces in both areas.

This was a hard section to write, because I know I will be called a "renewable basher." A friend of mine suggested that I never say anything bad about renewables, because "people will act as if you just shot their puppy."

As a matter of fact, I am in favor of renewables, when they are part of a plan. The job I wanted most in the world, and I eventually got, was Project Manager in the Renewable Resource Group at the Electric Power Research Institute. One of the things I liked best about renewables is that the entire fuel cycle is usually contained in one geographic area. The solar panel is on the roof and supplies energy to the house. The geothermal fluid (water, steam) comes out of the ground and spins a turbine, and then the cooled fluid is inserted into an old well to keep the underground field pressure up. There are no pipelines or coal trains for geothermal fluids.

However, renewables with a renewable portfolio standard in an RTO is a miserable combination. Solar panels that feed the grid on a hot summer day: great idea. But how many does the grid need, and how much will they raise the price of power, compared to traditional plants? Solar panels need to be matched with quick-acting resources such as gas turbines. Okay, how many gas turbines? When solar panels are added, how many more gas turbines need to be built? (Remember the NBER study that showed that you need 1.14 MW installed of quick-acting backup for every MW of installed renewables?) No RTO can limit solar, because they are supposed to be fuel neutral. All an RTO can do is ... set up another type of auction.

In the RTO areas, without any real oversight of requirements, renewable resources are overbuilt, and they lock the grid into gas-turbine backup. It's like building a car without a blueprint but taking the lowest bid for someone who can supply "part of a car." You may be able to procure wheels and tires, but transmissions will be way too expensive to buy. Buying tires may be inexpensive, but building a car needs a blueprint, and a car needs a transmission.

In other words, I am not against renewables per se. When I started working in the utility field, I aimed at expanding the role of renewables.

However, I am against greenwashing. "Come to our restaurant. It uses 100% renewable electricity. And the RECs were pretty cheap." I am against the no-planning idea that all types of generation are equivalent: that inexpensive, reliable baseload power is unimportant, because every part of the modern grid will be optimized to be "flexible."

The largest machine in the world is the North American grid. Not all parts on the grid are interchangeable. Not all types of plants have the same optimizations. With just a "market" for tires, a car won't be built. With a "market" that treats all sources of electricity as equivalent, despite differences in availability and other parameters, the grid becomes more fragile.

The grid will fail in the RTO areas. New types of auction carve-outs will not prevent rolling blackouts.

When that happens, in an RTO area, the buck will stop ... nowhere.

CHAPTER 36

OVERINVESTING IN RENEWABLES

In earlier chapters on renewables, I described research that showed that renewable mandates increase the cost of power and are a very costly way to cut greenhouse-gas emissions. Chapter 31, "RECS Are Us" described a New England cost-of-compliance study by James Bride of Energy Tariff Experts and a nationwide study of the costs of renewable portfolio standard mandates by Greenstone and Nash of the University of Chicago. In this section, we will start with the fact that renewables are expensive, that they cause problems on the grid, and that RECs and (to some extent) ZECs can be a shell game. Complying with state renewable standards is expensive, everywhere.

This chapter will dive a little deeper, however. We will look at the interaction between renewables and RTOs. I have not been able to find much research on this interaction, but I have found some information and come to some conclusions.

Basically, due to lack of accountability in the RTO areas, renewables can cause more problems (both physical and financial) for the grids in those areas. The tragedy of the commons has many forms.

Renewables in RTO and non-RTO areas

As I have described in several chapters, including the chapters on "RECs Are Us," renewables are expensive. And yet, in an RTO kWh auction, renewables can underbid other plants on the grid. This is not because the sun shines and the wind blows—for free. This is because renewables get considerable out-of-market funding, such as RECs and production tax credit. So, they underbid other types of power in the energy auctions, where the bids are "what is the marginal price of your next kWh produced."

Their marginal cost is often zero, because they get paid through subsidies. Despite underbidding other sources of power on an energy auction, the renewables add to the price of power for the consumer.

Since renewables can bid low on the energy auctions, renewable advocates in RTO areas can claim that renewables are the least-expensive thing on the grid. The fact that renewable mandates increase the price of power to the consumer is much harder to quantify than a "zero cents per kWh bid in the energy market." System costs are hard to explain. A study at the University of Chicago (or elsewhere) will not get the same publicity as a statement that "renewables are the cheapest power on the grid"—which usually means the lowest bid in an RTO auction.

One thing I have noticed is that the burden of paying for renewables seems to be higher in RTO than in non-RTO areas. This is not the same as saying that the air is cleaner in RTO areas. The RTO air may be cleaner, or it may not be cleaner. Many non-RTO areas have lots of hydro and often nuclear plants. Some RTO and non-RTO areas have significant percentages of coal on the grid. It's not an either-or situation about clean air.

However, in terms of renewables, states within RTOs seem to be more willing to encourage high percentages of renewables by a renewable portfolio standard, because their state PUCs do not have responsibility for the cost of power (the RTO auctions supposedly take care of that) or the reliability of power (the RTO auctions supposedly take care of that, too).

States in RTO areas can make renewable rules but do not have to take all the consequences of those rules. Renewable promoters in RTO areas can crow about their "low cost" on the grid, by which they mean their low bids into the energy auctions run by the RTO. Meanwhile, the fact that the consumer cost of electricity is rising does not seem to be something that anyone can change, in an RTO area.

In a regulated area, if your costs rise too quickly, a consumer group can appeal to the state Public Service Board (or equivalent). Pretty much, the buck stops there. The PUC has to answer for the cost rise and cannot push the blame onto the newest version of the Forward Capacity Auctions or equivalent. This view of the role of the PUC is overly simplistic, I know. But still, a higher percentage of the buck stops at the state PUC in a regulated area than stops at such any level in an RTO area.

The tragedy of the commons is a tragedy based on lack of accountability.

To encourage renewable generation, many states have set up rules for net metering. In net metering, an electricity customer sets up a PV (photovoltaic) system. Some states also include methane digesters or wind turbines in net metering, but solar photovoltaic (PV) is the most common type of renewable system that is net-metered.[220] The customer may be a homeowner, school, or a community system for an apartment building. The owner of a net-metered PV system (for example) doesn't use all the electricity the PV system makes. The homeowner will sell this electricity to the utility.

Here's the catch: in general, the utility must pay for the retail price for the net-metered power it purchases from the customer. This is sometimes described as "your electric meter runs backwards when you are selling your solar power to the utility." In other words, when you sell one kWh to the utility at retail rates, that erases the bill for 1 kWh that you bought from the utility at retail rates.

This is a bad deal for the utility. In most states, the utility can buy regular electricity on the grid for a few cents per kWh (let's say four cents), while a retail customer will pay the utility much more for a kWh of power delivered to the home (let's say eleven cents). In order for you to get this good deal by selling power through net-metering, your house has to be connected to the utility's system—which means that the utility has to have just as many wires and substations and overhead expenses as it did before you started net metering. But the net-metered customer is not paying for them.

The difference between the wholesale (grid price) and the retail (consumer price) is not a matter of the utility laughing all the way to the bank. For that price differential, the utility has to pay for distribution line maintenance, storm repairs, billing, help desks, providing electricity for low-income people who fall behind on their bills, paying the transmission authority for transmission costs. The utility must also pay the balancing authority and/or the RTO for balancing the grid and running the auctions. If the utility is meeting a renewable-energy requirement, such as a renewable portfolio standard, the utility must pay for RECs bought from other utilities. All these costs add up to the differential between the cost of wholesale cost of "raw" electricity in the kWh auctions and the cost of electricity as delivered to the customer.

In general, with net metering, the customer with a solar panel gets paid the retail price for the electricity he sells, instead of the wholesale price. In quite a few cases, the homeowner ends up selling so much power to the utility that the homeowner does not have to pay for power at all.

This is a very good deal for the homeowner. Too good a deal. In most states, not all customers are allowed to take part in net metering. There is usually a "cap," and net metering is "opened up for new sign-ups" and then closed again.

The reason is clear. If all kWhs of electricity were bought at the same price that it is sold to the customer—where is the utility going to get money for line maintenance and so forth?

Sometimes utilities get permission to ask homeowners to pay a connection fee for net metering. People fight this sort of fee, tooth and nail. As you can imagine, since early net-metering customers didn't have to pay such a fee, if a new customer has to pay it, that person may well say, "Why me? Why are you picking on me? My neighbor has no such fee." If fees are applied across the board, the neighbor, who never had to pay such a fee in the past, can well say, "Boy, bait and switch! There ought to be a law against this sort of thing. This is not what I signed up for."

And yet, if net metering were allowed for everyone, without a cap, we would have a "tragedy of the commons." Prices would rise for everyone. The "tragedy of the commons" is an expression for the consequences of overusing a resource. An area set aside for many people to graze their sheep is often called "a commons." It is an observed fact the number of animals on the land should be limited, to prevent the land from being overgrazed. With limits on the number of animals, the land can be productive for many years.

But the incentives for the person who owns some sheep is quite different from the plan that will lead to long-term productivity for the land. If one person puts a few extra sheep on the land, he makes more money than his neighbor because he had some extra sheep fattening up, and his neighbor didn't. When the overgrazed land eventually becomes too poor for grazing, the extra-sheep man has more money in the bank. This is the tragedy of the common land: it is easily exploited.

Similarly, if everyone is being paid the retail rate for a commodity (electricity) that comes with only wholesale attributes (selling plain kWh, no maintenance of distribution infrastructure), then the retail rate will become the new wholesale rate. By paying retail rates for wholesale power, there won't be enough money to maintain the distribution system. This is the equivalent of the commons being overgrazed. The price of electricity will rise, the utility will not be able to pay the new rates to everyone, the distribution system will be short-changed, and the entire system can fall apart.

A look at Nevada

Nevada is not in an RTO area.

In 2015, a Nevada legislator noted that net-metering customers received subsidies of $450 to $600 for selling their excess electricity to the grid. Other Nevada customers were paying for these costs. In 2015, the legislature passed a bill requiring the PUC to re-examine this cost shifting.

The PUC decided to pay net-metered customers the wholesale rates for electricity, rather than the retail rates. Existing owners were not "grandfathered" in at the higher rates.[221]

As you can imagine, this led to a huge pushback from net-metered solar owners. In 2016, in Wild West style, three Nevada solar owners tried to bring their guns to a hearing on net metering. They were turned away by security guards but said they would be back with their guns at the next hearing.[222] (They did not bring their guns back.) Most of the other protests against lowering the amount paid to solar owners were not nearly as dramatic. But the anger of the solar owners was sincere.

The results of all of this was a compromise in which rooftop solar owners are paid 95% of retail rates, but the payments can go down over time, to 75% of retail rates. It's a complicated law, because big solar installations get lower rates. In 2017, the new net-metering bill was signed into law.[223]

This is an ongoing story, and net-metering rules in Nevada are still in flux. However, Nevada can at least make its own determinations about net metering.

California also found itself with excessive costs due to net metering. It did not attempt to roll back the original payment plans, as Nevada attempted to do, though it did make the policy for new connections (NEM 2.0) stricter and more expensive.[224] A customer enrolled in NEM 2.0 is also automatically enrolled in "Time of Use" pricing. This means that solar panels which face southwest will be paid better than solar panels that face southeast. Also, under NEM 2.0, net-metering customers will have to pay non-bypassable charges on any electricity delivered by the utility. These charges fund energy-efficiency programs and low-income-support programs.

States and renewables and RTOs

In general, renewable portfolio standards and net metering are a state issue. In my opinion, the tragedy of the commons happens in the RTO areas. Let's face it, Nevada tried to keep utility rates reasonable by reining in net metering. And the state partially succeeded: net-metering rates have been lowered, and large systems can no longer make much money from net metering.

As far as I know, no state involved with an RTO has ever even attempted to roll back renewable subsidies. In the beginning of this chapter, I said that RTO rules are made with full understanding of the consequences. This is true for the rules that the RTO itself makes. However, at the state level, the RTO system allows a state to decide things like renewable portfolio mandates, and other states will endure many of the consequences of the original state's decision. The state may mandate high levels of renewables, but it will be up to the RTO to keep the grid stable.

Here in Vermont, we allowed Vermont Yankee to be closed down. The governor at the time actively campaigned for the plant to be closed. Vermont now buys 80% of its electricity from out of state.[225] The reliability of the grid is not our problem.

In a vertically integrated state, especially in a small state, the cost of purchasing power can be clearly coupled to the prices that the customer pays. In this case, there is pressure against excessive renewable mandates. In a non-RTO area, the state PUC is going to look at sources of electricity and consider reliability issues without the requirement to be "fuel neutral" about reliability.

In RTO areas, with their rules, tariffs, auctions, and wide geographical scope, some of those pressures seem to be relieved. The operative word here is "seem." The pressures of cost and reliability still exist, but in the RTO area, these are always someone else's problem.

CHAPTER 37

DISTRIBUTED GENERATION

MANY RENEWABLE ADVOCATES hope that the world will move away from big, central electricity generators, with their large switchyards and high-voltage transmission lines. Instead, they favor distributed generation, sometimes described as DER (distributed energy resources). These resources will be small generation installations, near the consumer, perhaps owned by the consumer, perhaps part of a microgrid.

Renewable advocates claim many advantages for this new type of grid, though they usually explain that the grid has to be modernized to be a "smart grid" before the transition can take place. The reason the grid needs to be modernized is that electricity production and consumption will be a two-way street. An electricity consumer will also be an electricity producer, and so communications between the grid and the "prosumer" (producer and consumer) will be important.

Becoming a prosumer

BEFORE GOING FURTHER INTO the pros and cons of the prosumer and the smart grid, we have to ask ourselves some questions about who these "prosumers" are. A Department of Energy webpage about prosumers and consumers shows a man receiving electricity and a bill from the power company. He looks very unhappy, holding his bill in his hand and frowning. He is merely a consumer.[226]

Figure 17: Consumer and prosumer (Department of Energy)

In the next panel, a cheerful woman has two-way communication with the electricity company: she receives electricity from the power company, and she also sells electricity to the power company. She sells electricity from her solar panels and from a wind turbine conveniently located near her house. She is a prosumer. She is not holding a bill.

There are some not-so-hidden problems with this graphic. First, let's look at what kinds of renewables lend themselves to prosumerism. Despite the cheerful woman with the wind turbine, very few wind turbines are owned by consumers. There's a reason they call it "Industrial

Wind" and install huge turbines on windy ridges and in the ocean. A small turbine in a non-windy area (where most people place their houses) will not make much power and will be a maintenance hassle. Few homeowners would choose to own a wind turbine, especially when compared to the relative ease of owning some solar panels.

Also, wind turbines can make "whooshing" noises and cause sunlight to flicker. If they are of any size (a home turbine may not require this), they also have to have a red light at the top to warn planes of their presence. Whether or not the noises, flickering sunlight, and lights from a turbine are a health hazard (a matter of debate), most people don't put one up right next to their house. As a matter of fact, in Falmouth, Massachusetts, neighbors have been successful in arranging for two local wind turbines to be removed.[227] Not everyone feels this way about wind turbines, but at this point, most people do not want the turbines too near their houses.

Are there other types of distributed generation that people can install, in order to be prosumers? Dairy farmers can have methane digesters and small diesels attached to the digester. They would use the manure from their barns as fuel. So dairy farmers can be prosumers.

What if I own a house and a woodlot? Maybe I could be a biomass prosumer?

Not likely. A prosumer doesn't just use biomass at home in a wood-burning stove. A prosumer interacts with the grid, supplying the grid with electricity and taking electricity from the grid. I would have to build a wood-fired boiler, raise steam, spin a turbine, attach a generator, and connect the whole thing to the grid. No, biomass electricity is not suitable for home use, and it's not likely to turn people into prosumers.

Owning a wood lot can save a person a lot of money on home heating and even water heating. But a prosumer must be in two-way connection with the grid. A college campus might be a prosumer with wood-fired electricity, but an individual will not be.

We could go down the list of other renewable-energy sources (waste burning, small hydro), and we will realize that some institutions may be able to host Distributed Energy Resources for the New Smart Grid. Most residential consumers will, at best, be able to have a solar panel on their house. In other words, for the individual consumer, Distributed Generation will look a great deal like Solar Net Metering.

Net metering means the person with the solar panels on his roof will be paid more than market rates for his power—a good deal for him. His neighbors, who do not have solar panels on their roofs, will have to pay the extra costs of their neighbor's power. Their electricity costs will rise—a bad deal for them.

So, let's go back to the graphic. We have a sad white male who merely buys power and a cheery woman of color who is a prosumer. Is this realistic? Well, of course, it is. A white male may or may not have solar panels on his home. The same is true for a woman of color.

However, while an individual woman of color may indeed have solar panels on her house, it has to be admitted that more people of color live in cities, often in apartment buildings, where installing solar panels is not very practical. "Community access" solar systems enable people in an apartment complex or area to share a larger solar installation. Still, there has to be a space for that installation. Solar installations are far more common in suburbia and rural areas than they are in cities.

The two cartoon characters have it backward. As a poster child for a consumer, we might show a woman of color with a city apartment house behind her. As the poster child for a prosumer, a white male standing in front of a lavish suburban home would be a better image.

By the way, I want to give the artist credit for getting her point across. In this cartoon, the existing grid is an expensive, old-fashioned patriarchy thing, while the future is two-way, inexpensive energy that includes women and people of color. The only problem is that the *reality* is almost the exact reverse of what the cartoon shows.

Utilities or the end of utilities

THERE ARE STATE AND utility programs, including loans and so forth, to help get low-income people set up with solar arrays and energy-saving devices. In a *New Yorker* article in 2015,[228] Bill McKibben describes how the Borkowskis, of Rutland, Vermont, a family of very modest income, improved their home-energy situation. McKibben is the founder of 350.org, an organization that campaigns against fossil fuels and hopes to stop climate change. With the help of various programs, the Borkowskis were able to get new insulation, electric heat pumps, and a small solar array. The Borkowskis remade the energy profile of their house, and there was no upfront cost to the family, thanks to Green Mountain Power's financing the improvements. As McKibben stands in the Borkowski kitchen and reviews their before-and-after-the-transformation electric bill, he describes himself as feeling a "fairly rare emotion: hope."

I also enjoy a story with a happy ending. Installing insulation is always a good thing, and if the state or the utility wants to lend the homeowner some of the costs, all the better. However, my thoughts about the Borkowskis as prosumers are not quite as cheerful. They at least own a stand-alone house in Vermont. I suspect it is no great shakes as a house, but it is not an urban apartment building.

As I described earlier, if everyone were to become a prosumer, the price of power on the grid would go up, because no business can sell its product to its customer (the prosumer) at the same price it buys its product from its supplier (the prosumer). In the *New Yorker* article, McKibben writes cheerfully of a utility "death spiral." Enough people may become prosumers, and then utilities—big, centralized utilities—may go out of business. Alternately, they may go into the business of leasing heat pumps or solar panels. Forward-looking utilities like Vermont's Green Mountain Power will try to "build a deep emotional and intellectual relationship" with their customers. In contrast, McKibben describes Arizona utilities as "scared" of such a change.

I see a different ending as more probable. I can imagine net-metering projects coming to a crashing halt. They are already restricted in most states. This would leave many prosumers in the lurch.

The end of net metering would leave some prosumers with financially useless solar panels. However, this is not the most negative ending I can imagine. The more-negative ending is that the old utility companies cease to exist as places that supply kWh and instead work at building emotional relationships with their customers.

At that point, the Borkowskis and others will have no source of power after the sun goes down. Conceivably, if the Borkowskis were well-to-do, they might have invested in a large solar installation and also in battery storage. Even so, their system will work best if it is not winter or cloudy.

In earlier chapters, I have written about the limitations of battery storage, and the greenwashing of batteries for "beat the peak." ("Beating the peak" is primarily about accounting, not energy savings.) For this section about net metering, I will simply note that, in general, people choose net metering because it is more flexible and cost-effective than investing in batteries.

If big utilities go out of business, someone in the neighborhood may purchase a set of noisy and polluting diesel generators. The generator owner can sell the power to their neighbors, through jury-rigged and dangerous wires. If you think I am being too pessimistic, well, maybe I am, for America. But I'm not making this scenario out of whole cloth.

In various parts of the world where the electricity supply is not reliable, people do sell to their neighbors. This is true in some parts of the Middle East. The generator owner is a power person in the area. (That's a bit of a cynical pun.) A recent *Wired* article on "Beirut's Electricity Brokers" describes the situation in areas where electricity is not provided by big companies but rather by freelance generator owners.[229] I am very glad that I don't live in Beirut.

For his important new book, *A Question of Power: Electricity and the Wealth of Nations,*[230] Robert Bryce visited Beirut and spoke with people there. They referred to the electricity "brokers" as the "electricity Mafia." They paid two electricity bills each month: one for about $35 to the state-owned power company, for the power they could provide, which was available about six hours a day. Then they pay around $100 a month to their local "mafia" generator. Bryce asked one man why he didn't just buy his own generator, since he was paying his neighbor a significant amount of money. The answer was that, if he broke away from the local "mafia" generator, he might be killed. At the very least, the wire to his generator would be cut. Bryce reports how a clash between two generator-owners left two people dead and required the Lebanese army to end the violence.

I am very glad that I don't live in Beirut. And I cannot imagine my neighbors in Vermont threatening me with death, even if they owned a generator. Still, it is worth noting what people will do, and what fears they will endure, in order to have access to reliable electricity.

Microgrids

YOU MIGHT CONSIDER the situation in Beirut to be a sort of microgrid. People do not rely on a central power station, and the electricity-generation owners are not huge corporations. Instead, they are your neighbors.

Somehow, however, I don't think Beirut is the model people want you to think about when they look forward to a world of microgrids.

Basically, microgrids have the same problem as other distributed generation. A set of solar panels does not a grid make. There has to be something else, some backup power. In the U.S., assuming those big corporations still run their plants on the grid, this is not much of a problem. I can buy part of a solar installation and tell my friends I am on a microgrid—and the electricity will go on, even at midnight, because I am also connected to the big grid.

When microgrids are introduced into areas that do not have a connection to the larger grid, they have mixed responses. Almost every medium-size actual island is basically a microgrid, and usually it runs with diesel generators. Most of us would not consider that method of making electricity a particularly ecological or noble way to make power. However, most islands do run on imported fossil fuels.

What if the islands had microgrids that included renewables? Since many islands are very sunny (especially in the Caribbean), this would definitely cut the use of fossil fuels. As I noted earlier, the problem with distributed generation is the backup. Backing up renewables with a diesel generator could work. You still need that diesel, though.

It's easy to say, "Baseload is an outmoded concept." It's not quite as easy to live in an area where solar panels don't provide reliable lighting at night. In India, solar panels and battery backup didn't do the job.

Around 2015, Greenpeace decided to put their money where their mouth was and provided a poor village in the Bihar region of India with solar panels and battery backup. I believe the plan behind this Greenpeace experiment was to show that their stand against fossil-fuel usage was not the same as a plan to deny electricity to poor people.

It didn't quite work out the way Greenpeace wanted it to work out. After spending $400,000 on the installation, they invited the former Prime Minister of the state of Bihar for the inauguration of the system. He was greeted by villagers who had lined up to protest: "We want real electricity, not fake electricity!"[231]

"Real" electricity comes from the grid, and the grid electricity in that part of India comes from coal. The first morning after the system was installed, the batteries were drained overnight. One young man had hoped to study in the early morning before he had to go to work in the fields. He discovered that his lights would not go on in the early morning. "We want real electricity!" People want the lights to go on. The title of Gayathri Vaidyanathan's article in *Scientific American* says it all: "Coal Trumps Solar in India."

If wealthy people feel that they need to encourage very poor people not to use coal—well, those wealthy people are going to have a problem. I would recommend that they watch Hans Rosling's wonderful video "The Magic Washing Machine," about the different types of energy poverty.[232] When you have enough electricity to run a washing machine, women become empowered. When you can run a washing machine, you have "real electricity." And you are probably connected to the grid.

Climate justice

So far, in America, our distributed generation is backed up by a robust electric grid, rather than by freelance brokers. I am happier than if I lived in Beirut, for sure. However, I need to say, for the record, that I dislike net metering intensely.

Net metering raises the price of electricity for those who do not have it, which is why many states have put caps on the amount of net metering allowed. Only some people can sign up for net metering. The rest of the people see the cost of net metering reflected in their own electric bills. I remember a friend of mine saying that "I love my net metering, Meredith. And you should, too. After all, you are paying for it." (He had worked with me at the Electric Power Research Institute, and he knew a good deal for himself when he saw it. He also liked to tease me when possible.)

This cap-on-net-metering system leads to anger between neighbors and even leads to conflict within families. In general, states allow net metering up to a certain percentage of electricity (the cap). At one point, Vermont raised the cap on net metering, so people who had not signed up in the earlier round could sign up. The rush to fill the open slots for "Look, no electric bills" was fast and furious, as you might expect. I heard one woman as she was berating her husband publicly for not being more Johnny-on-the-spot to get their family signed up for net metering. Did he have any idea how much his slowness had cost them? She is a very sweet woman, by the way, but the family was financially stressed. She could hardly contain her anger at how he had wasted their money by being slow.

Every time I hear about "climate justice" and how solar is going to bring climate justice to all of us, I think of that couple. Is it "climate justice" to race to beat your neighbor to a slot for low-priced electricity? Is this what the great new world of renewables will be all about?

CHAPTER 38

PERSONAL RESPONSIBILITY

This should be a workshop

I REMEMBER A MEETING of the ISO-NE Consumer Liaison Group in which a member of the audience made a comment that amazed me. In retrospect, I should have expected her remarks or something like them. But before I get to her remarks, I need to give some background.

I was a member of the Coordinating Committee for the Consumer Liaison Group for four years, and I am still a member of the group. In the CLG, we try to bring grid issues to the attention of the public, not just the "stakeholders" of the NEPOOL meetings. In general, we discussed some grid-level issue and usually included a speaker on the consumer implications of that issue. (We couldn't always get the speakers we wanted, since we couldn't even pay expenses for them.)

On the Consumer Liaison Group website,[233] you can scroll down to see the themes of some of our meetings. The meeting on December 6, 2018, was about how fuel-security concerns are changing the markets,[234] and September 20, 2018, was about exploring consumer choices for electrifying the heating sector.[235] The CLG places the PDF files of the slide presentations on their website, but it is not the same as going to the meeting, since the talk itself (and the questions) are not placed on the website at this time.

Some of the information presented at the CLG meetings is available in other ways. People can subscribe to *Utility Dive* or *RTO Insider.* However, these journals cover many topics and many areas of the country. Also, *RTO Insider*, an excellent source of information, costs more than $1000 a year.

In short, CLG meetings are an excellent way to find out what is going on in our local grid and to be able to ask questions of grid participants. On the Coordinating Committee, I helped set up the meetings. I consider my time on the Committee to have been a public service.

So, now we go back to the woman who was not impressed by our CLG meeting. Quite the opposite, as a matter of fact. She was angry. I will paraphrase what she said at the meeting and wrote on the after-meeting evaluation form:

> I can't believe this meeting! You are not concerned with environmental issues. You are not telling us what we can do to make the electric grid better. You should be running a workshop to show us how we can save electricity, install solar panels, insulate our houses, install heat pumps, and so on. Instead, we have a bunch of people talking about what their agencies or their companies are doing. This isn't an energy meeting. This is ridiculous!

Hubris

And there you have it. I am a great believer in taking personal responsibility, but I found her remarks discouraging. Too many people have the idea that their personal actions are paramount and that nothing else matters. It's a kind of hubris: *I will take these actions, and I will make the difference for climate.* Well, no. Even if everybody in Germany decided to lower their electricity use, insulate their homes, use LED lights, and so forth, Germany would still miss its climate goals.

The website *electricitymap.org* shows the grams of carbon dioxide produced per kWh of electricity. This site tracks many grids throughout the world in real time. Low-emission grids are shown in green, while higher-emission grids are shown in yellow, light brown, and dark brown. Germany and France use approximately the same amount of electricity per capita (6602 kWh per person per year in Germany, 6448 kWh per person in France, in 2016).[236] However, Germany has a high-fossil, high-renewables grid that emits about five times the carbon per kWh as France. German households that try to cut back on electricity use will not make German emissions anywhere near as low as the emissions of France.

Germans have chosen to cut back on nuclear and increase their use of coal. As you might expect, their carbon emissions are getting higher. Germany has said they will phase out coal by 2038.[237] Considering that Germany misses its climate targets on a routine basis, I personally wouldn't count on this coal phase-out. It's easy to announce a coal phase-out. Germany will have a hard time actually doing it. It will be virtually impossible for Germany to phase out coal, especially since they expect to close all their nuclear plants before they close their coal plants.

Comparing Germany and France shows that we make decisions as a society, and these decisions matter. Life is not just about personal virtue. Life is also about civic choices. Running a low-fossil grid is important. Knowing what it takes to run a low-fossil grid is important.

Of course, personal choices are always important. However, personal choices are not the only important issue in our energy future.

Getting back to the woman who was upset at our CLG meeting: a meeting can be a real energy meeting, even if it is not a workshop about insulating your house.

I am far from the first to notice that "personal responsibility" alone won't meet our climate goals. For one thing, the climate is worldwide, and the use of electricity is increasing worldwide. Joshua Goldstein and Staffan Qvist wrote an excellent book about worldwide energy use, *A Bright Future: How Some Countries Have Solved Climate Change and the Rest Can Follow.*[238] On page 49, they quote a report from the government of India that says that three quarters of electricity in India comes from coal, and that will not change significantly in the coming decades.

In *A Question of Power: Electricity and the Wealth of Nations*, Robert Bryce used a set of databases to divide countries of the world into three groups: Unplugged Countries, where per-capita electricity use is less than 1000 kWh per year, Low-Watt countries, where electricity use is between 1000 and 4000 kWh per year, and High-Watt countries, where electricity use exceeds 4000 kWh per year.[239] Bryce quotes studies that show that the Human Development Index (HDI) increases with increased electricity use—but does not increase with more use above 4000 kWh per capita. (The HDI includes life expectancy, education, safe water, and other measures of human health and happiness.) More than 3 billion people live in Unplugged Countries, including India. These people hope and plan to improve their lives by using more electricity and living lives with higher HDI levels. Their increased use of electricity will mostly depend on an increased use of coal. For example, India will not abandon coal and plans to use even more coal. Chapter 11 in Bryce's book is titled: "It's not possible to keep the lights on without coal."[240]

Looking at the worldwide growth in energy use, it is clear that I can put up some solar panels on my roof, but it would be better for the climate if I could encourage an international aid program that would help build nuclear power plants in India. In the final chapter of *A Bright Future,* Goldstein and Qvist suggest ways that countries can decarbonize their electric grid, as Sweden and France and Ontario have done, with a mixture of renewable and nuclear energy.

Perhaps the best example of why a little is a little, not a lot, is David JC MacKay's book, *Sustainable Energy—without the hot air.*[241] This classic book, published in 2009, is dedicated "to those who will not have the benefit of two billion years' accumulated energy reserves." MacKay does the calculations: if we captured every drop of water that fell in the English highlands and got it to run through a hydro plant, how much electricity could we make? He concludes that "if every river were dammed and every drop [of water] perfectly exploited," Britain could make only 1.5 kWh of electricity per person per day from hydro power.[242] One sixty-watt light bulb, running the entire day, uses 1.5 kWh of electricity.

MacKay (who died too young) was Regius Professor of Engineering at the University of Cambridge and Chief Scientific Advisor to the UK Department of Energy and Climate Change.[243] He wrote his book to encourage people to do the calculations and get beyond the hype that it merely takes willpower to be completely green.

In the context of "personal virtue as saving-the-world," the second section of MacKay's book is titled: "Making a difference." The first chapter of that section is titled: "Every BIG helps."[244] For example, he investigates the role of phone chargers on electricity requirements. Keeping them plugged in or unplugging them does not have a big effect. They use one-fourth of one percent of a home's electricity. As MacKay says: "The 'if-everyone' multiplying machine is a bad thing because it deflects people's attention toward 25 million minnows instead of 25 million sharks. The mantra 'Little changes can make a big difference' is bunkum when applied to climate change and power."

We have to make big changes in our power supply. We will have to use more electricity in order to reduce fossil-fuel use in the heating and transportation sectors. McKay hopes to get electricity from four sources: Britain's renewables, "clean coal" (if that is feasible), nuclear, and renewables from other countries. As MacKay says, we need to approach other countries "with great politeness" and make fair treaties to obtain their renewables. (He is obviously trying to avoid a new version of oil wars—desert sunshine wars.) MacKay hopes that solar power from other countries' deserts can be a source of renewable power for Britain.

Doing the numbers, as MacKay does, shows that Britain simply can't make it on renewables alone, and everybody "doing a little" would not be much help.

The same is true for many other countries. As a matter of fact, some friends and I did the numbers for replacing Vermont Yankee with renewables.[245] It wasn't going to work. Our report, *Vermont Electric Power in Transition,* is outdated but still mostly accurate. Vermont would require billions of dollars and tens of thousands of acres for the renewable installations. And Vermont would still need fossil backup since grid-level storage is not currently feasible.

We made a law! We're going to do it.

THERE ARE MANY LAWS and resolutions in U.S. statehouses such as "This state will get 80% or 100% of its electricity from renewable sources by (fill in the year)." In general, these resolutions make the unstated assumption that "by then" the backup problem will be solved. Batteries will save the day, perhaps, or a new type of renewable will be invented. And so forth. It reminds me of a cynical saying: "We don't just hope for miracles, we count on them."

Instead of looking at statehouse resolutions which may or may not lead to real-world results, we should look at what has worked to decarbonize other grids and follow those examples. If new types of carbon capture or new types of batteries can save the day, we can add them to our plans when they are available. It is not a winning strategy to make a resolution in a statehouse while ignoring techniques that have actually worked for other countries and regions.

In *A Bright Future*, Goldstein and Qvist show that countries that have solved climate change are using all their low-carbon resources: hydro, nuclear, renewables.[246] The governance of our RTO grids, however, is pushing them inexorably to being high-renewables, high-fossil, like Germany. This trend has to be talked about. Not every meeting about energy can be a workshop on how to insulate your home.

Before I leave this topic, however, I want to say that several states (Illinois and New Jersey for two) have passed laws encouraging low-carbon sources of all types: Renewables, hydro, and nuclear. Some states are ahead of the curve. I am heartened by these new plans, and I hope other states will follow.

CHAPTER 39

MANIPULATING THE CUSTOMER

Time-of-Use pricing

UTILITIES KNOW THAT IF customer demand did not vary so widely, their work would be easier and more profitable. They would not have to keep so many power plants ready to meet peak demand. They would not have to buy power when fuel and power prices have soared due to high demand. And, perhaps, their customers would not have to pay as much for power, because the utilities would need fewer power plants and would not have to stand ready to fork over "whatever it takes" to buy fuel or power when the prices rise.

Historically, many utilities have dealt with this issue by giving customers a choice in pricing, depending on whether the customer wanted to use power at peak or off-peak times. This is "time of day" or "time of use" pricing, and it is becoming more common in Europe and the United States. A 2017 article in *Commodities News*[247] shows that 90% of electricity customers in Oslo, Norway, buy power at spot-indexed prices. In Estonia, 70% of suppliers offer flexible pricing, and 20–25% of customers choose these contracts. In 2020, the European market is supposed to have a much higher percentage of this sort of flexible pricing.

In the United States, some utilities offer time-of-use pricing: for example, in Iowa, Alliant Energy allows customers to reduce their electricity bills by choosing to use power at the correct time.[248]

Unlike the "Beat the Peak" greenwashing described in chapter 24, "The Not-Stressed Grid in Summer," time-of-use pricing can be a win for the consumer *and* for the climate. It can be challenging for a utility because the revenue stream becomes somewhat unpredictable. However, as noted in my examples, some customers have the choice of using time-of-use pricing or another type of contract. Much residential electricity use is rather inflexible. People will turn on the lights when it gets dark, and they will make dinner at dinnertime. In consequence, some customers do not bother with time-of-use pricing, perhaps thinking that the cost savings will not be worth the hassle of being concerned with how expensive the electricity is right now.

Unfortunately, in the new world of renewables and RTOs, customers may not necessarily be given a choice. In Bakersfield, California, PG&E did not give people a choice.

The Bakersfield Effect

In 2012, Pacific Gas and Electric (PG&E) attempted to put smart meters in Bakersfield, California. Their plan was heavy-handed. PG&E instituted a program of time-of-day pricing, but they did not give people an opportunity to opt out. At least on one occasion, the smart meter installers were met with a gun-wielding woman who ordered them off her property. The protests spread. In other parts of

California, Californians barricaded roads, refusing to allow trucks carrying smart meters to pass.

Bakersfield customers who had smart meters installed often saw their bills soar unexpectedly. Electricity bills often went up threefold. Bakersfield is a very modest town in the southern part of California's central valley. It is not Palo Alto, where wealthy Silicon Valley entrepreneurs can watch cooling ocean fog rolling across the hills on many evenings. (When I first moved to the Palo Alto area, I was disappointed in the cool evenings. It was often too cool to sit outside, even in summer. I had grown up in the Midwest, where summer evenings were a true treat.) Unlike Palo Alto, Bakersfield is inland and hot. The people are generally not rich.

According to Gretchen Bakke's book *The Grid: The Fraying Wires Between Americans and Our Energy Future*,[249] PG&E blamed the weather and the people of Bakersfield for the rise in the electricity bills. The weather was warmer in July 2010 than it had been in July 2009, according to PG&E. Also, the people of Bakersfield had not gotten accustomed to the time-of-use pricing that had come in with the new meters. They hadn't adapted to the fact that electricity cost more in the afternoon than it had cost in the morning. Basically, PG&E was claiming that the high prices were the fault of the customers.

According to a recent article on smart meters for New Jersey,[250] PG&E eventually admitted that the meters they had installed in Bakersfield malfunctioned when they got too warm. In other words, PG&E's claim was incorrect. The large increase in Bakersfield electricity bills was not because the people of Bakersfield did know how to work with their smart meters. The increase was due to PG&E's malfunctioning smart meters. This type of incident does not make installing smart meters acceptable to the ratepayers.

This entire series of events is sometimes called "The Bakersfield Effect."

Smart meters

The Bakersfield Effect was more about utility arrogance than about smart meters or time-of-day pricing. But smart meters can be unpopular for other reasons, despite their many advantages.

Smart meters can be a great benefit to utilities and consumers. For one thing, they pinpoint outages automatically, and utility response can be faster. In the old days, the utilities had to wait for customers to call, and then (often) they had to wait for more customers to call. One customer calling in an outage didn't let the utility understand how extensive the outage was or the most likely place where a repair was needed. With smart meters, the utilities can send the repair crews to the right place a lot faster. Also, smart meters eliminate the jobs of meter readers. Your smart meter is read automatically.

Few people enjoy seeing jobs eliminated. Still, there is no question that smart meters, read automatically like E-ZPass® lines on Northeastern highways, save money for employers and save some hassle for the customer. No more fumbling for change while approaching the toll booth or hearing a meter reader walking on your property at unpredictable times. Like automatic readers at toll booths, smart meters are the future.

Yet many people do not like smart meters. As a strong supporter of personal privacy, I myself have some issues with smart meters. My concerns are not overwhelming. I have a smart meter in my home, though I easily could opt out. However, I do not consider my concerns to be trivial, either.

In Vermont, since smart meters save money for the utility, there were proposals that you should pay if you opt out of having a smart meter in your home. Without a smart meter in your home, utility costs for servicing your home will go up, and the idea is that it is unfair to make other customers pay such costs. Paying a fee to opt out of having a smart meter was first implemented in Maine.[251] However, the Vermont legislature passed a bill allowing Vermont utility customers to opt out of smart-meter installation for free.[252]

Very few people actually opt out of having a smart meter. But many people distrust them.

There are several reasons that people don't like smart meters. Some are concerned with problems I consider to be minor or non-existent.

One problem is that smart meters can interfere with some WiFi systems. In general, rearranging the system can solve this. In Maine, the utility is willing to consult with the homeowner. In some cases, the utility bought the homeowner new WiFi equipment.

Some people are concerned that smart-meter radiation will affect their health. However, a smart meter exposes you to far less electromagnetic radiation than a microwave or cell phone emits, as described well by Tom Evslin in an article in *VTDigger.*[253] In his article, he compares the exposure from a smart meter (4 microwatts per square centimeter for one-twentieth of a second every fifteen minutes) with the exposure from talking on a cell phone (1000–5000 microwatts per square centimeter, while you are talking.) Your microwave also exposes you to more radiation than a smart meter.

An important question, however, is whether the smart meter is optional. In the first smart-meter rollouts, people did not have the choice to opt out of receiving a smart meter. In contrast, people can choose whether or not to use a microwave or a cell phone. When people are forced to do something that they consider dangerous to their health (put a smart meter in their home), they will get angry. Hopefully, the ability to opt out of smart meters will allay this concern.

However, there are bigger issues—and harder-to-solve issues—based on privacy concerns and on control. The issues are related but not the same. I will start with privacy.

Privacy

A SMART METER CAN collect extensive data about your home. In a recent court decision before the Seventh Circuit Court of Appeals, the court concluded that the readings from the smart meters are a "warrantless search" (a search of your property without a search warrant) but that it was a "reasonable search." Part of the reason the smart meter was considered a "reasonable" search is that "the search is unrelated to law enforcement." The judge also wrote that the court's decision could change "if the data were more easily accessible to law enforcement or other city officials outside the utility." This subject is explored in more depth in an excellent *Utility Dive* article by Robert Walton.[254]

The smart-meter data can inform the utility when the home is likely to be vacant, what appliances are in the home, and when these appliances are in use. We can also look at smart meters from the utility point of view, specifically by reviewing a series of webinars made by Berkeley Lab in November 2018. The webinars are designed to help utilities use their smart-meter data.[255] The series explains how utilities can assign their customers to "segments" by load profile, and then assess whether each customer is a good candidate for certain types of time-of-day pricing. For example, if a customer already has a usage peak at a different time of day than the general utility peak, offering that customer time-of-day pricing may not save the utility any money. Basic and Advanced Customer Segmentation is covered, which includes grouping customers with similar load profiles.

All the customer-segmentation schemes described by the Berkeley Lab webinars are based on the customer load profile and how much it varies, day to day. At this time, utilities are not routinely obtaining data on individual appliances. Such non-intrusive load monitoring is a subject of academic research and start-ups such as EEme (recently acquired by Uplight, formerly known as Tendril).[256]

According to an article about smart meters in the website *northjersey.com*, in 2014, California passed a law forbidding utilities to share a customer's electric or gas usage with third parties without the customer's permission. (The utility still has to comply with subpoenas from law-enforcement agencies for such data.)[257] This protection is important because once data has been shared with a third party, it is no longer legally protected from further sharing.

In the future, another source of information about home use of appliances can be the use of Energy Star appliances. Some Energy Star appliances have "connected functionality," which basically means that they can communicate with the local utility. Not all Energy Star appliances have connected functionality, but Energy Star has set criteria for clothes washers, clothes dryers, dishwashers, lighting, refrigerator-freezers, room air conditioning, pool pumps, and thermostats. A list of appliances and Energy Star-connected functionality specifications dates can be found on a University of California-Irvine webpage.[258] Connected-functionality appliances can share information with the local utility, and they can be entered into a utility's demand-response program.

With "demand response" in that last sentence, we have moved quite smoothly from privacy concerns to control concerns. In other words, not only can the utility learn what appliances you are using and when you are using them, but if you enter those appliances into a demand-response program, the utility can turn them off.

Control

Demand-response utility programs have been around for a long time. They are usually a set of voluntary agreements that utilities make with large electricity users. Some large electricity users have been willing to cut their use of power when the utility asks them to do so. Since these large users are usually businesses, they do not cut their power (and therefore affect their own operations) as a charity to the utility. The utility generally pays demand-response users, either by paying directly for the cut in electricity usage when it happens or by giving the customer a discounted rate on all electricity use.

These programs are voluntary, and most big users do not subscribe. As I noted in the chapter on the Winter Reliability Program, ISO-NE asked large businesses to bid into a demand-response module of the 2017–2018 Winter Reliability Program. In this program, the businesses would be paid for using less power when the grid is stressed. ISO-NE didn't get many demand-response bids, because not many businesses want to use less power when the weather outside is below zero. As I wrote: "In 2017–2018, ISO-NE paid approximately $24 million for oil as part of the Winter Reliability Program but paid only $34,000 for demand-response bidders."

ISO-NE spent more than 700 times as much on oil as on demand response, because they got so few demand-response bidders.

Demand response may not be popular with businesses, but it is very popular with some environmentalists—the ones who favor less energy use and favor paying for "negawatts." (Armory Lovins introduced the "negawatt" concept in 1985. It basically consists of setting economic incentives for using less electricity.)

If you have connected Energy Star appliances and a smart meter, you have the equipment to be enrolled in a demand-response program. The big question is whether joining this program would be voluntary on your part or whether the utility would simply enroll you. Customers cannot be blamed for thinking that the next step after the smart meter might be the utility controlling their electricity use—demand response.

However, it would be very unusual for residential users to be forcibly enrolled in demand-response programs. There are many voluntary programs and more such programs all the time.[259] Many of these programs are carefully constructed to give choice to the consumer. For example, the Southern California Edison Summer Discount Plan[260] allows people a choice of programs. A customer can sign up for a "maximum savings" demand-response program, in which the utility chooses when to turn off their air conditioning. Alternately, the customer can sign up for a program in which they save less money, but they can override the utility's demand-response control and keep their air conditioner running—a good system, in my opinion.

For many people, the idea that the power company can manipulate electric devices within their home is extremely distasteful. Can they turn off my air conditioner, delay the start of a cycle of my dishwasher, delay the start of my washing machine? For many people, this is unacceptable. In my blog post "The Oversold Smart Grid: Dismissing the Work of Women,"[261] I took aim at a local commercial extolling the smart grid. The commercial showed a dryer running in the middle of the night. As a woman who raised two children, I can tell you that laundry doesn't do itself. If that dryer is running, someone turned it on. And someone will have to be around to take the clothes out so they won't get wrinkled.

In my opinion, installation of smart meters and enrollment in demand-response programs should always be voluntary.

Manipulating the customer

SINCE 2008, ELECTRICITY USE in the United States has been fairly flat[262] and has actually declined about 6% on a per-capita basis.[263]

And yet there is a constant push, especially in RTO areas, to urge customers to conserve, to time their electricity use to off-peak hours, to allow the electric companies to manipulate energy within their home.

Gretchen Bakke's *The Grid* notes that this push is new:[264]

> This interest in controlling customer use is brand-new. For the whole previous history of the grid, the utilities did what was necessary to adapt to our whims, whether greedy or conservative.... This is why utility intervention "feels" so weird ... We are rightly suspicious of them, even if spuriously so.

There are many loaded words in that section. We customers must change both our usage and our attitudes. For example, we can no longer expect utilities to attend to our whims.

I don't mean to be too harsh about the Bakke book. It includes enlightening and well-written information on the history of the grid. In this book, I am telling a different story about the grid. I look specifically at the RTOs and how their auctions are leading us to a high-carbon, expensive grid.

I always encourage anyone who is interested in the future of our grid to visit the website *electricitymap.org.* This is a real-time map of the grams of CO_2 per kWh of various grids, worldwide. Green areas are low-CO_2 grids. Yellow and brown areas are high CO_2 grids. This real-time site shows what is really happening on the grid: where it is clean and where it is dirty (at least with regard to CO_2). France and Sweden, with nuclear energy and hydro power, are green. Germany, with its renewables plus fossil backup, is not green. Germany is high carbon compared to France. The RTO rules push us in the direction of Germany.

The colors on this map have almost nothing to do with choices made by individuals. It is very tempting to believe that, if we stop using electricity whenever we want to use it, and instead we make personal sacrifices, everything will be all right. It won't be. We merely need to look at the high-emissions German grid to understand that endpoint. An individual German making a decision about using a clothes dryer or air-drying their laundry will not lower the country's carbon emissions nearly as effectively as if Germany made a different choice about energy choices. At this point, however, Germany is closing nuclear plants, opening coal plants, and promising to close the coal plants sometime in the future. The German grid emits much more carbon dioxide than the French grid, and the decisions of individual German consumers won't change that.

We need to make decisions as a society, not just as individuals. But will these decisions be easy? In this chapter on "Manipulating the Customer," we have reviewed consumer choices. Before we look at a plan for the future, it is time to look at the choices energy suppliers are making.

CHAPTER 40

LEAVING THE RTO AREAS

As Gretchen Bakke noted in her book, utility interest in controlling the customer is a relatively new phenomenon. Utilities used to cater to the "whims" of the customer. Now, many utilities, especially in RTO areas, are focusing on controlling customer use of electricity.

In this chapter, however, we are going to look at the RTO areas from the point of view of the generators, not the customers. Many of the utilities that own generating units are making a decision about the RTO areas. They are leaving.

In the Jump Ball chapters, we looked at the way that one RTO area had to twist itself in knots, with "jump ball" filings to FERC and complex and contentious formulas for rewards and punishments ... in order to keep the lights on in bad weather in the Northeast.

In this chapter, we will describe how the very ordinary operation of the so-called "energy markets" assures that no power plants will make money, except for subsidized plants that get money from sources other than selling electricity or electricity services. This is the tyranny of the RTO system. Utilities leave the system, unless they can arrange to get subsidies.

Choosing to leave

A SEQUENCE OF EVENTS began to happen among the power plants in RTO areas. It wasn't visible to most people, but it was happening. As nuclear plants and coal plants threatened closure (or did close) in the RTO areas, it was generally assumed that the problem was that they could not compete with newly inexpensive natural gas and, to some extent, with renewables. Therefore, these closures were taken as showing that "the market is working." Close observers of the market could note that this story was incomplete, but for most people, this was the story. "Just can't compete."

For example, by 2017, Entergy had closed Vermont Yankee in Vermont and announced it would close Pilgrim in Massachusetts and Palisades in Michigan.[265] Everyone thought that this meant that Entergy no longer wanted to own nuclear power plants in RTO areas. (Entergy also announced that it would close Indian Point in New York State, which was probably the endgame of a long fight by anti-nuclear organizations to force Entergy to build cooling towers for the plant.) So, at that point, it was assumed that the anti-nuclear forces and the market had successfully closed the "outmoded" nuclear plants, at least the ones owned by Entergy in the RTO areas.

What was not noticed in this general discussion of the troubles of the nuclear plants is that Entergy had also announced it was selling a gas-fired plant in Rhode Island.[266] Entergy was removing itself from the RTO areas. Entergy's choices weren't fuel-specific: they were RTO-specific. But, mostly, people did not notice.

Nuclear plants are considered to be "exceptional," and coal plants are considered to be "exceptional" in other ways. Therefore, it was easy to miss the real story: "Entergy exits RTO areas."

People did begin to notice the real RTO story when FirstEnergy announced, quite explicitly, that it was leaving the RTO areas. In late 2016, FirstEnergy CEO Charles E. Jones spoke at the Edison Electric Institute, saying that FirstEnergy was planning a 12- to 18-month timeline to implement their transition: "Transforming to a Regulated Company."[267] Will Davis wrote a thoughtful post about this in ANS Nuclear Cafe, and he included an analysis of the implications of Jones's talk.[268] Among other things, FirstEnergy was considering shutting down large nuclear plants, plants that had economies of scale and that had been widely considered to be profitable. The company said that it needed a "regulation-like" structure to support the nuclear and coal plants.

In my own post a few days later,[269] I noted that FirstEnergy was following the Entergy model and leaving the RTO areas.

Time has passed since 2016, and as I write this, FirstEnergy has announced shutdown plans for three large nuclear plants in Ohio and Pennsylvania.[270] Some of the Ohio plants have received subsidies.[271] FirstEnergy also asked the Department of Energy to provide subsidies of some type to keep its Pennsylvania nuclear and coal plants open but has not received any subsidy as of this writing. FirstEnergy also plans to close its 1,300 MW coal station in West Virginia.[272]

Okay, so the nuclear and coal plants are in trouble, because of low-priced natural gas. And Entergy decided to sell a natural-gas plant for some reason. There weren't articles and commentaries about the trend of leaving the RTO world. The general feeling was "natural-gas prices are down, so, of course, things happen."

Ummm ... well ... in that case ... why did Exelon decide to close the Mystic Station gas-fired plant in Massachusetts?[273] The major reason is that they are losing money on that power plant. Now, Mystic Station gets much of its gas from LNG, as well as gas from pipelines. This makes the Mystic fuel supply more expensive than perhaps other gas-fired plants. On the other hand, it is a gas-fired plant, and it is available in miserable weather (or hot weather) when the price on the grid is high. If it can't turn a profit, things are fairly bad.

RTOs and the missing money

As I noted in an earlier chapter, RTO areas make many rules. They have more tariffs than vertically integrated utility areas. Often, each new rule comes with a new auction: for capacity, for winter reliability, for capacity of "sponsored resources." And each new rule is "fuel-neutral" even when it isn't. As described in the Jump Ball chapters, the rule in the New England RTO for Pay for Performance is intended to reward plants for being online in fiercely cold weather, when the grid is stressed. But the rule doesn't pay coal, oil, and hydro plants, because, for Pete's sake, everyone expects them to be online in cold weather, and you can carry this fuel-neutral stuff only so far. (End sarcasm.) To be fair, the ISO-NE plan would have paid them, but the NEPOOL participants considered such payments to be a waste of money. They made their own plan, which FERC adopted.

RTOs are not a market. If you are raising zucchini, you could factor into your prices the fact you bought good riverside land at high prices, and you have a tremendous yield of zucchini. Meanwhile, someone with rather bad land that yields few zucchini per acre, but the land was cheap, can base his prices on his situation. RTOs, however, force everyone to bid into (usually ... some RTOs are different) two main markets: the energy market, which you can consider to be the marginal price of the next zucchini (you have to bid low, because your land is very productive, and the RTO is watching you and knows everything about your excellent soil), and the capacity market, which you can consider to be the fixed costs of land and capital. In the capacity market, you probably will have to take the clearing price set by the guy who bought really awful, hilly, somewhat acidic soil and struggles to grow even the next zucchini.

Since neither of you can recapture your major costs, you and the hilly-land guy are both going to be hard-pressed to make a profit.

In the utility business, this has been described as the "Search for the Missing Money." All the utilities are missing money in the RTO markets, whatever kind of power plants they have.

Entergy left the RTO markets first. Entergy had mostly nuclear plants, and shutting such plants includes permissions from the local grid and all sorts of dealings with the Nuclear Regulatory Commission. It is taking Entergy a long time to get out of the RTO areas. FirstEnergy is following a similar path. Meanwhile, Exelon is closing a major gas-fired plant. I personally think this is just the opening wedge of an exodus. If companies can't find the missing money to run their plants, they will stop running them.

Small update: Exelon received reliability payments from the grid operator to keep Mystic (the gas-fired plant) operating. The fate of that plant past 2024 is unclear.[274]

When I am really cynical, I think that the plants in the RTO areas are playing a kind of a game of musical chairs. In the long run, somebody is going to have to provide power for the current RTO areas, and, to be sustainable, they will have to make money by doing so. Using 100% renewables is probably not ever possible and is certainly not just around the corner. Traditional power plants will have to keep running. Every time Exelon or Entergy or FirstEnergy shuts down a plant, prices on the grid tend to rise (less supply, no change in demand).

As prices on the grid rise, the companies that have kept their plants will find that those plants are more profitable.

The prices will go up, even if the cost of fuel remains low. And the endless drama of the RTO areas will continue.

Throughout this book, I have looked at the complexities of the tariffs in the RTO areas. With very complex systems, it is sometimes hard to see whether a consequence is actually the *reason* for a rule or if it is an unintended *consequence*. Is the result a feature or a bug? I believe that there are few "unintended consequences" in the RTO areas. In other words, whatever happens is a "feature."

Yes, I am aware that this last statement (the rules made at RTOs are "features," not "bugs") may seem somewhat at odds with some of my earlier assertions. For example, at the end of chapter 31, "RECs Are Us," I said that I agreed with Annette Smith. She had observed that the laws about RECs were becoming so complicated that legislators could not predict the effects of the laws they had just passed. Have I reversed myself now by saying there are few "unintended" consequences?

No, I haven't really reversed myself. Laws about RECs are state laws. State legislators are not RTO Participant Committee members.

State legislators tend to be part-timers (the legislature is not always in session) and often have day jobs separate from their government roles. They also rarely have a fraction of the staff they would need in order to keep up with all the issues they face. In contrast, in an RTO, Participant Committee members usually have clear guidance and good support from their home companies. They can actively support tariffs that will have good consequences for their employers. And there seems to be plenty of money for analysis.

RTO analysis may be time-constrained in an RTO system: "You must file with FERC by a certain day." However, in my opinion,

RTO analysis is rarely seriously cost-constrained. If the Participants Committee wants a lengthy and expensive analysis, it usually gets one. That is my observation, anyway. For example, chapter 19 described the history of the "Fuel Security Report." ISO-NE ran twenty-three scenarios in their first Fuel Security report. Then they ran another hundred scenarios after the Participants Committee objected to their first batch of scenarios.

In my opinion, the results attained by the RTO tariffs are features, not bugs.

When you go to ISO-NE headquarters, or you look at the websites of the other RTOs and the POOLs and FERC, or you go to a CLG meeting, in which lawyers and consultants are among the speakers, you can't help but realize that all the consequences are discussed, modeled, and discussed again, from the economic and legal points of view. In any jurisdiction, there are tens of people engaged full time in such analysis. If there is a consequence to a change in a rule or tariff, it has been considered, lobbied for, lobbied against, and decided upon.

Most of the people discussing the policy change are either "stakeholders" (insiders) or consultants to the "stakeholders." In my opinion, all these people think through the consequences and try to influence them. These pressure groups have skin in the game, and they are watching and influencing the game.

And yes, a lot of this happens behind closed doors. There are no sunshine laws on the grid.

I also recognize that, since I am not welcome behind most of those closed doors, when I write that a certain consequence was predicted and planned for, I am exceeding my actual knowledge. I may come close to "ascribing motivations," which is always unwise. However, I do have a certain level of understanding about the people involved in making these rules. Ascribing motivations can be unwise, but it is also unwise to think that all the lawyers, economists, and executives involved in the tariffs are not very bright. It is unwise to think these people constantly stand around saying, "Gosh, who would have thought of that?"

Who wins? Who loses?

When power plants leave an RTO area, it is worth looking at which part of the payment stream has been recently affected by price changes within the RTO. In general, in an RTO area, renewable plants are assured payment and will continue to operate. Single-cycle gas plants will fight hard to keep capacity payments high (see the MOPR and CASPR controversies, above) but will usually continue to operate, though some of them may well leave the area. (Entergy and FirstEnergy, as discussed earlier.) Coal and nuclear, dependent on energy payments, are likely to close, as are combined-cycle gas plants.

In general, in the RTO areas, single-cycle gas-fired plants (not combined cycle) plus renewables will ultimately be used for everything: covering baseload, following load, covering peak load, and providing reserve. As noted above, gas is just-in-time, and renewables are not dispatchable. The RTO areas are inexorably sliding toward complete dependence on gas deliveries for the reliability of their power.

James Bride showed this trend toward gas in a Consumer Liaison Group presentation in December 2016.[275] The electricity bill of a typical customer in Connecticut decreased from 2008 to 2016: the total amount decreased from $121 to $110. However, the components of the bill changed: transmission, distribution, and renewable-energy charges went from around $52 of a $121 bill to $75 of a $110 bill. As Bride wrote: "Low natural-gas prices have helped ameliorate consumer cost pressures." The non-energy prices grew, but the lower prices for gas made up for this.

If gas prices rise

LOOKING AT BRIDE'S CHARTS, we realize that if the gas prices rise again, consumer bills would soar. Consumer bills are now loaded with much higher transmission, distribution, and renewable-certificate payments than they were in 2008. If gas prices go up, could those payments continue? Or, as Bride wrote in his list of "difficult questions" based on these changes: Does policy leeway evaporate if natural-gas prices recover?

I think policy leeway would indeed evaporate. In many RTO areas, including the Northeast, low natural-gas prices have offset the increased charges due to complex auctions and hidden subsidies for renewables. If and when natural-gas prices go up, prices will rise, rather suddenly, in the RTO areas.

Now that we have toured the grid, where do we go from here?

In the next section, I will try to answer that question. How can we set up grid governance that will lead to a reliable, inexpensive, and clean grid? Our first order of business will be the moral question raised by part of the Bakke book:

Do consumers have a right to use electricity?

From there, we will move to how we, as consumers, can influence policies on the grid. It isn't going to be easy.

The grid we want

BEFORE WE CAN DESIGN the grid of the future, we have to ask ourselves some practical and moral questions. The two big questions are:

- Do we deserve to have electricity available at all times? Or is an intermittent, fragile grid good enough?

- How important are the various aspects of our electric supply? That is, what are the values we assign to things like reliability, cost, low environmental impact, and low carbon-dioxide release?

IS THERE A WAY FORWARD?

CHAPTER 41

RELIABLE ELECTRICITY

Electricity when we want it?

At the end of the last chapter, I asked two philosophical questions about the electric grid. The first was about grid reliability, and the second was our choices for electricity supply.

My answer to the first question is simple."Yes, we deserve to have electricity available when we want it."

This may look like a simple case of "Meredith's Opinion." But it isn't. Energy poverty hurts people, and it hurts women and children the most. They are the ones inside the house, and they are the ones breathing pollution from the stoves and kerosene lamps that provide heat and light within the house.

When prosperity increases, energy use increases. Increased energy use itself increases prosperity.

As shown in Bob Hargraves's book *Thorium: Energy Cheaper Than Coal*, there's a strong inverse correlation between prosperity and birth rate. The more prosperity, the lower the number of children per woman.[276]

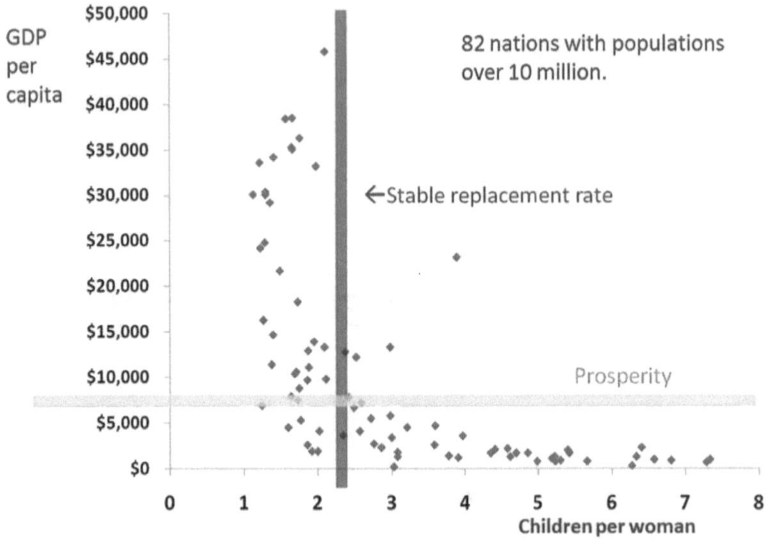

Figure 18: Births per woman vs GDP per capita (Robert Hargraves)

Reliable electricity also increases choices and health in men's lives. Remember the young man in India who wanted to study in the early-morning hours before he had to go to the fields? He wanted a better life for himself, and good lighting (energy use) was going to help him achieve it. A fragile grid (sorry, no electricity early in the morning) could not help him achieve his goals. (See chapter 37 on Distributed Generation.)

In his amazing book, *Factfulness*, Hans Rosling and his son and daughter-in-law trace the various levels of poverty and energy poverty. It shows how people's lives get better as prosperity increases.[277] Bill Gates said about this book: "One of the most important books I have ever read—an indispensable guide for thinking clearly about the world."

Steady electricity releases women from grinding labor. In his book *A Question of Power: Electricity and the Wealth of Nations*, Robert Bryce describes how electricity improves the life of women. Without electricity, women must pump water, heat water over a fire or on a stove, use the heated the water for washing or cooking, and constantly tend the fire or the stove. Bryce describes Lyndon Johnson campaigning for himself and for electricity in the hill country of Texas in the 1930s. With electricity, Johnson said, women would have water pumps and washing machines instead of washboards. They would have a way to keep cooked food overnight in the refrigerator. A woman's life would not be a constant round of drudgery. Lyndon Johnson was correct. Today, electrification in India achieves similar gains in release from drudgery.[278]

I know that some people think we should not have electricity whenever we want to use it. These people usually have a rosy view of a low-electricity-use lifestyle. In this view, lack of reliable electricity means that we will have more time to sit around the woodstove and chat with our family.

That is not what happens. When the grid is fragile and electricity is unreliable, the reality is less bucolic. To escape grinding labor, people seek a reliable source of power. Fossil-fired small generators are the usual remedy of choice when grid electricity is unreliable. Generators are popular, and not only in Beirut. In America, many people own their own generators. Generators tend to be noisy and polluting. However, when the choice is endless, repetitive labor or access to a generator, people usually choose the generator.

We need a reliable grid—a grid that provides electricity when we need to use it. A grid that is not fragile.

However, one question might be: "Who are '*we*'?" Maybe people will be making our own power. There's a lot of talk about microgrids and prosumers and making your own power. As I have shown in various earlier chapters, only a few of the electricity consumers in the United States can participate as prosumers.

Also, when people in America talk about microgrids, they usually mean "some solar, and something else, too." As Greenpeace discovered in India, solar plus some batteries doesn't meet people's needs, even for a poor village in Bihar. In America, "I'm in a microgrid" is usually some variation of a statement such as "I am participating in community solar, and I use grid power most of the time." In that sense, a microgrid can work—with the big utility grid as backup. It works because it is not really a stand-alone microgrid. The utility grid provides the reliable power.

A microgrid can work, if it is backed up by a utility grid or a sufficient number of fossil-fired generators. The end result needs to be reliable power.

CHAPTER 42

A HIGH-QUALITY ELECTRIC GRID

A high-quality electric grid must provide reliable power: electricity when we want it, as described in the previous chapter. However, there are other aspects of a high-quality grid.

A checklist for a high-quality grid

Here are my criteria for an electric grid that meets the needs of humans and is respectful of its effects on the environment:

1. The grid should work very reliably for all customers. Everyone should have access to energy, every hour of every day.
2. The power plants on the grid should be as clean as reasonably possible.
3. Similarly, safety concerns for nuclear energy have to be balanced with the positive benefits of the technology.

4. Electricity prices should be as low as reasonably possible. In particular, no residential customer should pay an increased bill in order to provide lower prices for another residential customer.
5. Low-carbon, non-fossil sources of electricity should be encouraged, as much as reasonably possible.
6. We should be ready to use more electricity, not less. If we want to reduce pollution from the heating sector and the transportation sector, we will have to use electricity in those sectors.
7. While there is much excitement about microgrids, solar power, and so forth, the grid design should acknowledge that only a small portion of electricity users will be prosumers.

Reviewing the list

YES, THERE ARE TWO problems with my list: The word "reasonable" and the issues around carbon dioxide.

First, there's that word "reasonable." What is reasonable to one person may not seem reasonable to another.

Let's look at fossil fuels, biomass, and other combustion sources of power. In terms of power plants on the grid, nobody wants unabated, high-particulate-release coal burning, or the release of large quantities of nitrogen oxides from unabated gas-fired plants. On the other hand, constant increases in pollution-control requirements will drive some types of plants off the grid as "uneconomical." That is often the purpose of those requirements. That is, increased requirements for one technology are often supported by the lobbyists of rival technologies.

In terms of nuclear energy, the constant increase in safety requirements for nuclear have the same effect as constant increases in pollution-control requirements for fossil plants. The plants become "uneconomical," and the rival types of plants get closer to being a monopoly. Nuclear plants are already very safe. No American has ever been killed by emissions from a commercial nuclear plant. (About fifty people were killed in Ukraine.) As one of my friends said, "How much lower than zero deaths do you want the nuclear industry to go?"

Second, some people may disagree about the necessity of using electricity that emits very little carbon dioxide (point five). I think low-carbon electricity is important for our future. If you don't agree, the other concerns should still be important.

The priorities for the future in this section are for the United States and other economically developed countries. In much of the world, simple access to reliable electricity is the main concern. For excellent ideas and descriptions of the priorities and choices for poorer countries, I recommend the books: *Thorium: Energy Cheaper Than Coal*, *Factfulness: Ten Reasons We're Wrong About the World—and Why Things Are Better Than You Think*, and *A Question of Power: Electricity and the Wealth of Nations*.

My personal preference

Each area of the country and each area of the world has its own constraints. Some areas have mountains that are good for hydro, and some don't. Some have easily accessible natural gas, and some don't. I hesitate to describe my preferences for a grid, because my preferences will not be suitable for everywhere in the world. However, I am also concerned that people will say, "She's great at describing the problem, but she doesn't have a solution, does she?"

First, I want to say that the problem I am describing is grid governance that is leading to a fragile grid. I encourage people to be aware of grid decisions and take action. However, in fairness to my readers, I will also answer the question "What would be my personal preference for a grid?"

My preference would be for a grid with nuclear plants running as baseload. They are non-polluting and very reliable. When running as baseload, they are also relatively inexpensive. A wind turbine may seem less expensive per kWh, but that doesn't count the system-wide costs of wind and solar, including their natural-gas backup. Nuclear plants are intrinsically more reliable than natural gas plants, because they have fuel stored on site, unlike just-in-time natural gas. So, I would look at a baseload of nuclear. I would also point out that baseload is a very high proportion of the total electricity demand, as described in the section "What is baseload?" in chapter 26.

For load following, hydro is the best, in my opinion. It starts quickly, and it is non-polluting. However, not all areas have hydro, and many areas that have hydro are reluctant to expand hydro, because dams have large effects on the local ecology. So, if the area does not have hydro, I would suggest a load following with a mix of smaller nuclear, plus wind, solar, and natural-gas backup. To keep a clean grid, however, it is important that these technologies be used along with the nuclear baseload. They should not be allowed to undercut the solid performance of the baseload plants.

Those are my preferences. This book is about the grid as a whole, and about how the RTO areas lead to grid fragility. Grid planning is about power plants, but it is also about location and power-line rights-of-way and other constraints. It is also best for a grid to have either some gas plants or some hydro plants. Gas and hydro plants are best for Black Start, that is, for restarting the grid when the whole area has had an outage. Yes, there are many constraints on designing a grid! Integrated Resource Planning: take it seriously!

I can't possibly describe the best grid for every area, but I have described the kind of choices I would make, and why I would make those choices. Every geographical area has its own constraints, but a local adaptation of these choices will lead to a clean and reliable grid.

CHAPTER 43

FIX THE GRID AND END THE DRAMA

Toward a better grid

IN THE PREVIOUS CHAPTER, I described a list of changes that would make the grid more robust.

In the RTO areas, the grid is being moved inexorably to a strong reliance on intermittent renewables, coupled with an equally strong reliance on just-in-time natural-gas delivery as backup. Several different scenarios could cause this system to collapse. Natural gas, the cure-all fuel for the grid and for houses, could become scarce or expensive. A long, cold windless period or a long, hot, humid period could overwhelm the natural-gas supply for electric generation. Wind generation is often hundreds of miles from the load center, adding another level of vulnerability to the transmission system. And so forth.

The question is: What shall we do about this? For the purposes of this chapter, I am going to assume that the reader has looked at my seven criteria for a high-quality grid (from the last chapter) and wants to know how to move toward implementing at least some of those changes.

First, we must acknowledge that the fixes will be difficult. No matter how necessary such changes are, the ratepayers have very little chance of implementing them. We are at the bottom of the pecking order. When I went to a luxurious breakfast meeting of NEPOOL, I was painfully aware that I was not considered a stakeholder of the grid. Stakeholders buy memberships in NEPOOL. Mere ratepayers are not stakeholders. I felt more like a serf than a customer.

Grid governance

ONCE SERFS HAVE DECIDED what changes they would like to see on the grid, the question becomes what type of grid governance would bring these changes. In my opinion, the older, vertically integrated utility model was better (despite its flaws) than the current RTO model. My issues with the RTO model include:

- Excessive tariffs and complex regulation.
- Excessive deference to stakeholders/insiders.
- Nobody has responsibility for keeping the grid operational.
- A grid that becomes more fragile with increased reliance on just-in-time gas without new pipelines to bring more gas.
- Extra consumer costs introduced by the RTOs, as described in chapter 13, "The Death of the Market Hope."
- States can make rules while assuming that some other state in the RTO area will bear the cost.

Travis Kavulla, who thinks the auction methodology can be a good thing, acknowledges some of these problems in his article.[279]

> [The] value proposition [of the auctions] collapses if their design includes unnecessary features that have the practical effect of putting a regulator's finger back on the dial, in place of firms making business decisions.

That is a sentence which pretty much describes the auctions as they are today.

Instead of looking at regulators and commentators, let's look at states. Margaret Thatcher, Britain's famous conservative Prime Minister, once said that "the trouble with socialism is that, eventually, you run out of other people's money."

To some extent, the California RTO is a poster child for how not to run a grid. California is closing down zero-emission nuclear plants, setting high requirements for widespread use of renewables, depending heavily on natural gas (no surprise there) and on imported electricity. California rates are far higher than they should be for a state with significant hydro power and in-state natural-gas supplies. But the California ISO is running out of California money.

Or rather, they are running out of California grid. California has provided such extensive supports for intermittent renewables that their grid is often overloaded. In that case, the renewables are often "curtailed." That is, they are not allowed to put power on the lines because the lines cannot accept so much power at once. Figure 19 shows the curtailment in California in the years 2018 and 2019.[280] One solution that California tries to propose is a regional grid, an RTO that covers more of the West. New transmission lines (probably) would send California renewables all over the West, within a SuperISO made up of California and neighboring states that are now vertically integrated. As you can see in figure 19, in most months, CAISO has to curtail tens of thousands of MWh of renewable power, because there is too much power for their system in the middle of the day.

Now, some people might think it was foolish and wasteful to build such an oversupply in the first place, but California's preferred remedy is to "regionalize" their RTO to the neighboring states. (A vertically integrated state system would have held a PUC hearing where citizens could have objected to the cost of the oversupply.)

Proponents of CAISO regionalization claim that it would allow "greater renewables integration"[281] in the Western states. California opponents of regionalization are concerned that California might begin importing power from out-of-state coal generators. After all, the oversupply is not a constant oversupply. Who knows what might happen when California renewables are not in oversupply?

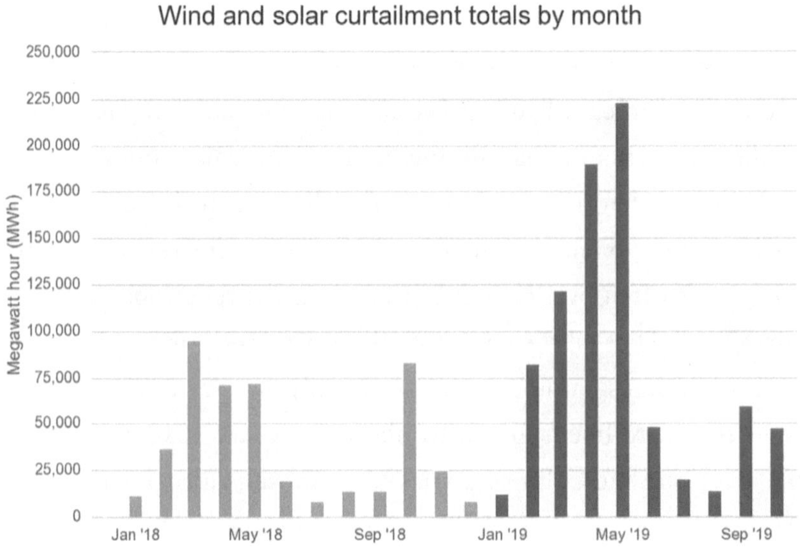

Figure 19: Wind and solar curtailment totals by month (CAISO)

David Roberts of Vox wrote about this in an article, "California's huge energy decision: Link its grid to its neighbors, or stay autonomous?"[282] I think the title of the Roberts' piece shows how most people (and probably Californians) think of the decision. It's California's decision to make.

Actually, it isn't. The neighbors have something to say about it, too. I remember when I lived in California, there were signs in Oregon: "Welcome to Oregon. Now go home." Then there were the bumper stickers: "Don't Californicate Oregon." The neighbors aren't always happy with the big rich kid on the block.

So far, even in the California statehouse, a bill to regionalize the California grid has gotten stalled, year after year. Also, if one looks at Nevada and how it has rolled back net-metering (see the discussion of Nevada in chapter 36, "Overinvesting in Renewables"), we can see that states that control their own policies are more responsive to the cost implications of those policies than states that must defer to multi-state RTO auctions.

In short, states that are not in RTO areas are usually quite happy with their vertically integrated utilities, local (state-level) regulatory oversight, and low rates. They're not in much of a rush to join RTOs.

CHAPTER 44

A BRIEF LOOK AT ONTARIO

The clean-grid RTO

IN THE BOOK *A Bright Future*, Joshua Goldstein and Staffan Qvist look at two grids that are low carbon and have reliable, inexpensive power. The two grids are France and Ontario. The Ontario grid is operated by an RTO: IESO, the Independent Electricity System Operator.[283]

As we move forward with ideas on fixing the grid, it is worthwhile to look at Ontario. How can an RTO system encourage a clean and reliable grid?

While acknowledging that every RTO has its own rules, there are still features that make the Ontario system quite different from other RTOs.

Some of the salient features of the Ontario RTO:

1. **One State:** The Ontario RTO is completely within the province of Ontario and is responsive to the needs and wishes of the

elected provincial government. In this aspect, IESO is similar to ERCOT, CAISO, and NYISO. Being responsible within one state helps the grid governance, but it's not sufficient to define a clean grid.

2. **Global Adjustments:** While the grid has energy auctions, generators obtain most of their funding from "Global Adjustments," not from auctions for the next kWh of electricity. The Global Adjustment price is set by a combination of long-term contracts, regulated rates for nuclear and hydro generators, and cost of conservation programs.[284] The Ontario grid has some characteristics similar to a system regulated by a state PUC. It is a one-state RTO, and the Global Adjustments take into account most of the fixed cost of the generation, not just the marginal cost. This brings the RTO closer to a grid with a "rate of return." In contrast, U.S. RTOs try to be "all-auctions-all-the-time." The 2019 Hourly Ontario Energy Price (HOEP), set by the energy auctions, was 1.83 cents per kWh, while the Ontario Global Adjustment price was 10.8 cents per kWh.[285]
3. **No Capacity Auctions:** IESO does not have capacity auctions at this time, though they are considering implementing such auctions. Since my local grid has capacity auctions, I am painfully aware of how capacity auctions encourage generators that have high marginal prices and low rates of participation on the grid. In other words, capacity auctions are a gift to peaker plants and a burden to baseload plants. However, because of the Global Adjustments payments, Ontario has different issues. Ontario is considering capacity auctions in order to encourage new capacity to come online, hopefully, at lower prices than those set earlier in power-purchase agreements.

Ontario's decisions

The Ontario RTO provides funding to operate nuclear plants, hydro plants, gas turbines, wind turbines, solar PV, and some bio-energy plants. It also has responsibility for central planning and procurement. The supply mix is set with input from the provincial government and from IESO planners. Energy-policy goals include (1) reliable electricity, (2) affordable electricity, (3) low-emission electricity, and (4) support for strategic energy investments in renewables, storage, natural gas, and nuclear.[286]

In support of these goals, Ontario has phased out coal-fired electricity generation. As their planning site says: "[this was] the largest single greenhouse-gas emissions reduction action in North America."[287]

Ontario's Global Adjustment payments support low-emissions generation. In contrast, other RTO energy markets require bids that are the cost of the "next unit of production," called the "marginal cost." The marginal cost cannot include capital costs.

Since low-emission plants have low or zero fuel costs, being paid the marginal cost for the next kWh is too little money. Marginal cost energy payments cannot fully support the operation of these plants. And these plants are needed in order to meet the grid requirements for reliability. In U.S. RTOs, the energy markets are the major funding source for many plants, especially baseload plants. (This funding problem is described further in chapter 15, "Selling kWh is a Losing Game.")

In contrast to marginal-cost payments, here is how the IESO site describes the Global Adjustment payment:

> The global adjustment (GA) is the component that covers the cost of building new electricity infrastructure in the province, maintaining existing resources, as well as providing conservation and demand-management programs.[288]

Let's look at the cost structures for various types of plants in Ontario.

- The major expense for an operating gas turbine is buying the gas. Therefore, the cost of the next unit of production can provide a considerable portion of that plant's financial requirements. Only 6% of Ontario's electricity comes from natural gas.
- A nuclear plant's major expense is its capital cost and employees. Ontario has CANDU nuclear plants: their fuel is natural uranium, which is far cheaper than enriched uranium.
- For wind turbines and solar PV, the major expense is capital cost. The marginal energy cost (land rent and maintenance) is low.
- Ontario has regulations that prohibit the use of coal for electricity production. This is a prerogative of state-level decision-making. Within a state, there is usually no requirement to be "fuel-neutral" about generating sources.

Meeting the goals, controlling the dispatch

To ensure that reliability goals are met, all plants in Ontario get most of their revenue not in the energy market but from Global Adjustment payments, based on their contracted energy price, less any fixed-cost contributions received in the wholesale energy market. In other words, their primary payment is not based on the marginal cost of the next kWh. The system is a hybrid market combining the integrated utility "rate of return" concept and the more usual RTO energy-auctions concept.

In this system, nuclear plants and run-of-the-river hydro plants run baseload. Their power is cut back only when there is insufficient demand on the system for even their baseload output. Gas, flexible hydroelectric, wind, and solar are cut back first. Flexible hydroelectric are hydro plants that have provisions for the grid operators to remotely control their output.

How does this baseload strategy actually work? "Floor prices" are the key. Ontario's "floor price" market rules prohibit wind and solar PV from offering energy at negative prices that force nuclear offline. Negative pricing by plants that get subsidies distorts markets: "I'll pay you to take my power" can be a good strategy for a plant that gets other subsidies, but it is a very bad strategy for a reliable grid.

In Ontario, wind and large solar PV facilities are forced to curtail before nuclear curtails. The wind and solar facilities directly connected to the IESO high-voltage grid are affected this way. In general, rooftop solar and small-scale, ground-mounted solar are not affected, because they are connected to the distribution system and not controlled by central dispatching algorithms.

Nuclear energy supplies more than 60% of Ontario's electricity.[289] Before the floor prices were introduced into market rules, curtailment of nuclear by wind and solar was a frequent occurrence. But, *floor prices*? Really? A first impression is that setting a price floor would cost money for consumers. It doesn't.

If negative pricing by wind and solar forced nuclear plants offline, those plants would be slow to ramp up again, and a multi-day nuclear outage could result. In the subsequent hours/days, while nuclear was offline, wind and solar would probably be available only part of the time. Natural-gas plants would need to start up. This would result in both higher emissions and higher energy costs for consumers.

Before floor prices were adopted, there were many meetings and presentations about the probable effects of implementing this system. A 2012 IESO analysis[290] estimated that floor-price rules would save consumers around $180 to $225 million per year and reduce emissions

1.6 to 2 million tonnes CO_2. These numbers were projected for the year 2014. Similar savings could be expected in subsequent years.

Under the older rules, without price floors, wind turbines would force nuclear units offline, with considerable cost to customers and increased environmental emissions (from the gas plants that took over when the wind died down). Under the new floor-price rules, the wind turbines are now curtailed frequently. Among other things, this means that the fixed costs of wind turbines have led to higher Global Adjustment payments for the entire grid.

Almost heaven?

ONTARIO HAS A LOW-CARBON, low-cost, and very reliable grid. The nuclear plants are the backbone of the grid, and coal is banned. Floor prices protect the clean baseload and also save money for consumers. The mixed system (Global Adjustments for fixed capacity costs, plus energy auctions) works well for consumers, cleanliness, and reliability.

Does this mean that, if other RTOs follow IESO's lead, all will be well?

Clearly, floor prices will improve grid stability if you already have nuclear plants available. Other things, such as outlawing coal, can be done only in a one-state or one-province RTO. Multi-state RTOs cannot pick technologies so explicitly.

A 2019 report by the Ontario Society of Professional Engineers,[291] notes that 10.2 TWh of emission-free power was curtailed in 2017. (1 TWh is a million MWh). This is enough electricity to supply all the power for more than a million homes. The report suggests types of pricing that would allow consumers to capture those TWh to decarbonize other sectors of energy use, such as heating and transportation.[292]

What is past is prologue, and what is past is legacy. Sometimes an area builds coal plants and then phases them out. Sometimes an area overbuilds wind turbines and then curtails their output. In both cases, there is wasted effort and money. In the Ontario situation, thoughtful retail electricity pricing could perhaps lead to less curtailment and an even cleaner province.

There is no method of running an electric grid that leads to a heavenly outcome, with everybody satisfied. However, if you look at tradeoffs and results, Ontario's hybrid market-system RTO looks very good. It may not have achieved the optimum grid (whatever that is), but the Ontario grid is clean, reliable, and affordable. The Ontario grid is basically a win for its residents and businesses. The Ontario grid is not a simple RTO that is run only by marginal-price auctions.

CHAPTER 45

WHAT WE CAN DO

What works to build a solid, reliable, clean grid

As a general rule, local control of the local grid works best. Yes, we need a regional grid, a balancing authority, and transmission lines. But RTO areas also have a proliferation of different types of auctions and FERC filings. These are unnecessary, expensive, and counter-productive in terms of building a resilient grid.

In my opinion, the vertically integrated utility market is a far better plan than the RTO system.

Would any state in an RTO area attempt to scale back net-metering, as Nevada did? I don't think so. State PUCs have many faults. And yes, they are susceptible to regulatory capture by the big utilities, and citizen groups must watch out for that problem. But state PUCs are also much more attuned to the requirements and the costs of the people in the state than an RTO is. State PUCs don't assume that the neighbors will take up the slack if they fail.

Indeed, I would like to see RTOs actually be what their name says they are: "Regional Transmission Operators." Right now, they are grid rulers who are not even accountable for the grid being able to keep the lights on. Under the control of RTOs, grid after grid is sliding toward overbuilt, curtailed renewables and just-in-time gas backup. In other words, the RTO grids are becoming more expensive and more fragile. And the buck never stops with them.

Local citizen groups

I BELIEVE THAT STEADY grassroots advocacy is the way to change society. Grassroots advocacy at a state PUC is comparatively easy. PUCs often have open hearings and take comments from the public. Tangling with the well-heeled crowds at an RTO Participants Committee meeting is quite another story, if the committee even allows non-insiders to attend the meetings.

Keep track of your grid

I CANNOT SHOW A TRUE road map for how to fix the endless slide into just-in-time gas at the RTO areas, but I have some suggestions.

- Subscribe to *Utility Dive* and other online resources (utility-trade magazines), and try to find out what is going on at your RTO.
- Follow your RTO's tweets, and subscribe to any newsletter that they publish for non-insiders. These are mostly on the web. Yes, these sources are mostly Public Relations, but after watching them for a few months, you will be able to get a glimpse of the reality behind the words.
- Go to your local state PUC meetings. Your PUC should announce these meetings on their website.

- Make public comments at PUC meetings or on their website, if they take comments there.
- You can't do much about what the RTO does, but by participating in your PUC, you can strengthen its ability to stand up to the RTO.
- Follow your local grid (your RTO will generally publish some information on the website), and write letters to the editor and op-eds. Bring grid governance to the attention of the general public.
- It's best to organize a small group, perhaps with a name like "Connecticut Grid Watch" as an informal group, or as a formal, organized non-profit. You need to talk to people about the grid. It's important to have a group and to trade ideas. Organizing as a group also gives the members a title to put with their names when they write a letter to the editor (Member of Connecticut Grid Watch) and so on.
- If your group is well-funded, a subscription to the trade journal *RTO Insider* would be worthwhile. It's expensive, but if nothing else, we all owe *RTO Insider* some thanks for suing to open the NEPOOL meetings to reporters and the public.

It can be difficult to find out what is going on at your local grid. But I suggest you try. Very few people at the insider meetings have your best interests at heart. You are merely a ratepayer. You won't have any influence on the grid unless you work at it.

FERC: Act nationally

At some level, FERC would seem even harder to influence than the regional RTOs. After all, the Commissioners at FERC are appointed by the president and confirmed by the senate. Looking at FERC, one might be forgiven for thinking about the old Russian saying: "God is too high, and the Czar is too far."

However, FERC dockets are on the web. FERC often asks for comments before they have rulings. I believe they actually read the comments. Frankly, comparatively few people comment. You can be one of those people. I think an individual has more potential influence with FERC than with an RTO.

Still, the FERC website is immense. I would start by looking at *Utility Dive* and so forth to discover what FERC dockets and rule-makings are coming down the pike and which are relevant to your area of the country.

FERC may be a thicket of lawyers, and it sometimes holds closed meetings when necessary to protect proprietary information. However, FERC has nothing that corresponds to the many insider-only meetings at an RTO.

If you are not in an RTO area

Congratulations! Don't move to a different state.

However, you still have to watch the grid. A grid-watch group and attendance at PUC meetings are still important. This book started with the shortcomings of the old, integrated utilities. Those problems still exist. Luckily, you have more influence within your state than you would in an RTO, but even so, you must watch out for regulatory capture and lightly disguised insider deals. Follow your PUC, as suggested above for the RTO areas.

Without citizen awareness and participation, no system will work right.

The slide to natural gas

I think that everyone should be aware of the tendency of our grids to slide toward being completely natural gas. The problem is that natural gas is just-in-time delivery. This doesn't matter too much if natural gas makes 25% of the power on the grid, but it matters a lot when it makes 50% of the power, as is happening in the Northeast.

Few people push directly for an all-natural-gas grid. However, unless abundant quick-responding hydro is available, the imposition of high renewable portfolio standards means that the grid needs more and more natural gas as backup. In watching the grid, try to be aware of the reality of how unreliable renewables are on the grid. They accelerate the slide to all-gas, all the time.

Do not be fooled by the idea that a high renewable percentage is the most virtuous form of grid. It isn't—not when the renewables have to be backed up with fossil.

In my opinion, behind every group pushing for impossibly hard to meet renewable standards for a state or region, there's another well-funded group eager to sell even more natural gas.

Energy Secretary Rick Perry was mocked for his concern for fuel stored on-site, but it wasn't and isn't a trivial concern.

For those who are concerned with climate change, as I am, let me just note that natural gas is a fossil fuel, and nuclear energy has zero emissions. Nuclear energy is also the champion for fuel stored on site, 18 months worth of fuel at most plants. So, you can be in favor of low greenhouse gases and also be in favor of the security advantages of having fuel stored on-site.

It is easy to dismiss New England's problems as New England's problems. But they are the direct result of overbuilding renewables, closing nuclear and coal plants, moving to natural gas to back up the renewables, and not building new gas pipelines. This is a scenario for grid fragility. If gas just-in-time doesn't make it just in time, then the ratepayers will suffer the rolling blackouts. This progression to rolling blackouts can happen (and is beginning to happen) in other RTO grids.

The RTO areas encourage their grids' progression to fragility, because, at the end of the day, they are not responsible for grid reliability.

In an RTO area, ensuring reliability is a buck that never stops. As ratepayers, unwelcome at most of the RTO meetings, we need to start paying attention.

We need to pay attention

There are systematic problems with grid governance in the RTO areas. The grid is becoming more single-fuel, more vulnerable, and more expensive. Insiders make the rules, and the public cannot participate in a meaningful way.

In trying to talk about the grid problems, I have found that most people have deep opinions on power plants. They hate coal or nuclear, or they hate wind turbines, or whatever. They are often completely unaware of how the ratepayer-serfs are getting robbed by the insiders in an RTO area. Grid governance just isn't on most people's radar.

I hope you will pay attention to grid governance and take part in the debates. Ratepayers are not usually "allowed" to take part in the debates, but we have to try.

If we are not concerned with the grid, we will not have a safe and happy country to leave to our children.

I can't say it any more starkly than that. We must take the grid away from the insiders, or our children may be outsiders, in some very unpleasant ways.

ACKNOWLEDGMENTS

Many people helped me in the multi-year task that became this book. I want to thank my husband, George Angwin, who advised me, encouraged me, reminded me that my work is important, and was amazingly kind to a sometimes-frantic wife. He followed that activity by doing huge amounts of work on many areas of the book. For example, it is his doing that the endnotes are basically in *Chicago Manual of Style* format. In this regard, I also want to thank Linda Nelson and Margaret Leslie, who helped with similar issues at various stages of the book preparation.

I owe a tremendous debt to my "first readers" of the early manuscript: Rod Adams of *Atomic Insights* blog, and Margot de l'Etoile, retired from administration of the Dickey Center at Dartmouth College. Their critiques and insights improved this book. George Angwin was another important first reader. If I continue to mention George in every capacity that he helped, however, I will never be done with this section.

Others helped in various ways. Some helped by encouraging me, some by answering my questions, and some by reading sections of the book. These people include Jarrett Adams of FullOn Communications, Edward Warren, formerly of Combustion Engineering, David Schumacher, director of the film *The New Fire*, James Bride of Energy Tariff Experts, and Ken Alton, former manager of the hydro plants on the Connecticut River. I thank Howard Shaffer and Willem Post for being my long-time energy friends, and thanks to Donn Dears of *Power for USA* blog for our many thoughtful discussions. Thank you to three authors who were inspirations to me, and who were also kind enough to read this book and give me very helpful advice: Robert Bryce, *A Question of Power: Electricity and the Wealth of Nations*, Robert Hargraves, *Thorium, Energy Cheaper than Coal*, Joshua Goldstein, *A Bright Future: How Some Countries Have Solved Climate Change and the Rest Can Follow.* Thank you to Elizabeth Wilson, executive director of the Irving Institute of Dartmouth, for discussions and for the huge *Energy Law and Policy* book which she was kind enough to give me. I had other helpful conversations with Armond Cohen of the Clean Air Task Force, and Jesse Jenkins (now of Princeton). I also appreciate Warren Van Wyck for his careful analysis of graphics about the New England grid, and Eric Hittinger of Rochester Institute of Technology for his great bicycle analogy about VARs. For the Ontario section, I owe great thanks for the guidance from Paul Acchione, P.E. Fellow of the Canadian Academy of Engineering, and for input from Canadian bloggers Steve Aplin (*Canadian Energy Issues*) and Scott Luft (*Cold Air Online.*) I thank Dan Nott of the Institute for Cartoon Studies. He shared an early draft of his book with me: a forthcoming illustrated book on Infrastructure, *Hidden Systems*, to be published by Random House Graphic. I thank everyone who was on the Coordinating Committee of the Consumer Liaison Group of ISO-NE at the time I was on the Committee, and all the people from ISO-NE who coordinated and cooperated with that committee. Another thanks to Mathijs Beckers, Tim Maloney, and Mike Conley for their careful work rebutting overly enthusiastic plans for 100% renewable power.

In this section, I have thanked some of the numerous people who were helpful to me. "Helpful to me" does not necessarily mean "agree with me." I want to make it clear that many of the people whom I thank do not agree with some or all of my conclusions about the RTO systems. As a matter of fact, there are others who shared information with me but did not want their names used in conjunction with the book. I thank them for their input ... you know who you are.

I also need to thank people for their courtesy and kindness. In this book, I share some harsh opinions of grid-level decision making. However, I want to make it very clear that the people in NEPOOL, ISO-NE, and other organizations are courteous and kind, even to someone like me who disagrees with many of their decisions. RTO governance is not made up of people who are constantly rubbing their hands together and saying, "I'll get you, my pretty, and your little dog, too!" The problems arise due to the way the RTOs are set up. The problem is the rules, not the people.

I am grateful to 1106 Design for their help with putting this book together and making it into something that is clear and professional in its design and editing.

I am grateful to everyone who helped me. Thank you for your help.

Of course, all mistakes and errors in this book are fully my own.

GLOSSARY

AESO: Alberta Electric System Operator.

Alberta Electric System Operator (AESO): Regional Transmission Organization for the Province of Alberta, Canada.

ancillary services: Services that ensure the reliability of and support for the transmission of electricity to serve load. Such services include frequency-response services (regulation or automatic generator control), spinning reserve, non-spinning reserve, replacement reserve, reactive and voltage control.

ampere (amp): A unit of electric-current flow.

BA: Balancing Authority.

Balancing Authority (BA): The entity responsible for integrating resource plans for a given area (the balancing authority area). These responsibilities include integrating resource plants ahead of time and maintaining in real time the area's load-resource balance and interconnection frequency. Balancing authorities are responsible for an assigned area and for complying with NERC standards applicable to BAs.

baseload: The minimum load on the system at all times. Baseload is the electricity required at the time when minimum electricity is required. For most systems, that would be electricity required around 3 a.m. on a mild-weather day.

baseload generating unit: A generating unit used to satisfy all or part of the minimum load of the system. Such units generally produce electrical energy continuously and at a constant rate.

CAISO: California ISO.

California ISO (CAISO). Regional Transmission Organization for most of California and parts of Nevada.

capacity: The continuous load-carrying ability, expressed in megawatts or megavolt-amperes, of generation, transmission, or other electrical equipment. For generating equipment, this is usually the same as "nameplate capacity," which is the size of the equipment (in MW) on the manufacturer's nameplate.

Capacity Market: See Forward Capacity Market.

capacity factor: The ratio of the electrical energy a generating unit produces for a certain period of time to the electrical energy it could have produced at full operation during the same period. For example, a 400 MW steam-fired power plant with a capacity factor of 80% can produce an average of 320 MW over an extended period. A wind farm rated at 50 MW might have a capacity factor of 30%, so it can produce an average of 15 MW over an extended period of time.

capacity value: A somewhat controversial concept concerning how the availability of intermittent renewables interacts with the energy needs of the power grid. Discussed in the section of this book on various amendments to the capacity markets (MOPR and CASPR).

CASPR: Competitive Auctions with Sponsored Policy Resources.

Competitive Auctions with Sponsored Policy Resources (CASPR): In the ISO-NE area, the CASPR auction is a separate Forward Capacity Auction for generators that have significant out-of-market funding (such as wind turbines that receive production tax credits and can sell Renewable Energy Certificates). If such generators bid into the usual Forward Capacity Markets, they would lower the overall clearing price. Discussed in the section of the book on amendments to the Capacity Markets (MOPR and CASPR).

CC: Combined Cycle.

clearing price: When the quantity of resources (energy, capacity) meets the requirements for resources, the highest-cost resource utilized will set the clearing price. All the other resources also receive this price. This is the general rule of economic dispatch (see below), but it is not universally followed. For example, the RTO can rule that a certain plant (or type of plant) must be a "price taker," not a "price maker." That plant would not be allowed to set the clearing price.

For an example: If the grid operator needs 300 MW over the next hour, and plant A bids in 100 MW at $10/MWh, plant B bids in 100 MW at $15/MWh, plant C bids in 100 MW at $20/MWh, and plant D bids in 100 MW at $25/MWh, plants A, B, and C will be selected to meet the 300 MW demand. Plant C will set the clearing price at $20/MWh, and plants A, B, and C will all receive the clearing price. In this example, plant C is the price maker, and plants A and B are price takers.

combined cycle (CC): A type of plant (usually gas-fired) where the waste heat from a combustion turbine is used to raise steam in a steam cycle. These plants are more efficient than single-cycle combustion turbines.

CONE: Cost of New Entry.

contingency: The technical word used for a grid problem, meaning the failure of a power plant, transmission line, or substation.

Cost of New Entry (CONE): In ISO-NE markets, the price of capacity ($/kW-month) that is required to attract new capacity. CONE is designed to attract lowest-capital-cost new capacity, such as gas turbines.

curtailment: In the utility world, curtailment is the act of reducing or restricting energy delivery from a generator to the electrical grid. The grid operator curtails a generator for various reasons, often including potential overloading of transmission lines.

Day-Ahead Energy Market: The market that produces financially binding schedules for production of electricity one day before the operating day. It is based on calculations of probable electricity demand throughout the day.

demand resource: An entity receiving payments for lowering the demand for power by, for example, shutting off equipment that uses power.

dispatch: The action of a control-room operator issuing electronic or verbal instructions to generators, transmission facilities, and other market participants to start up, shut down, raise or lower generation, and so forth.

distributed generation: Generation provided by relatively small installations directly connected to distribution facilities or retail-customer facilities. A rooftop solar photovoltaic system is an example of distributed generation.

distribution: The delivery of electricity to end users via low-voltage electric power lines (typically less than 69 kV). It can also mean the transfer of electricity from high-voltage transmission lines to lower-voltage lines.

dual-fuel capability: The flexibility and storage capacity of a generator to use other fuels as well as natural gas. Payments for storing oil at dual-fuel units has been the main strategy of ISO-NE Winter Reliability Programs.

economic dispatch: Selecting generation resources in order of cost, with the least-expensive bids selected first. This is the usual type of dispatch, which attempts to meet the electricity demand as inexpensively as possible. This is also sometimes called "dispatch in merit order."

electric energy: The ability of an electric current to produce work, such as heat, light, or mechanical energy. Electric energy is measured in kilowatt-hours (kWh), megawatt-hours (MWh), or gigawatt-hours (GWh).

electric power: The rate at which electric energy is transferred or used to do work, measured in kilowatts, watts, or megawatts. This is also the basic measure used in capacity markets, which are generally auctions for electric-power availability.

electricity market: A system for buying and selling electricity, generally using auctions to set the price. Electricity markets generally include electric-energy markets (for example, price per MWh), capacity markets (for example, price per MW that is available for use during a given month), and ancillary-services markets (for example, price for frequency-regulation services or for fast-start reserves).

Electric Reliability Council of Texas (ERCOT): The Regional Transmission Organization whose territory covers most of Texas.

ERCOT: Electric Reliability Council of Texas.

fast-start resource: A generation unit that can start up and be at full load in less than 30 minutes. See also non-spinning reserve and spinning reserve.

FCA: Forward Capacity Auction.

FCM: Forward Capacity Market.

Federal Energy Regulatory Commission (FERC): The Federal Energy Regulatory Commission is an independent agency that regulates the interstate transmission of electricity, natural gas, and oil. FERC also regulates the transmission and wholesale sales of electricity in interstate commerce. FERC issues Orders (such as FERC 1000, which regulates some aspects of transmissions). FERC also must approve major tariff changes made in the interstate market. In the latter case, it rules to approve or disapprove changes suggested by RTOs and other entities.

FERC: Federal Energy Regulatory Commission.

Forward Capacity Auction (FCA): See Forward Capacity Market.

Forward Capacity Market (FCM): In New England, the market in which ISO-NE predicts the needs of the power system three years in advance. It then holds an annual auction (FCA) to purchase capacity resources to satisfy the region's future needs. The FCM is designed to send appropriate price signals to attract new investment and maintain existing resources where they are needed, thus ensuring the reliability of the New England electricity grid. Similar markets exist in other jurisdictions. The FCM rules are often tweaked to accomplish various goals. See MOPR, CASPR, and Pay for Performance.

frequency: The rate of oscillation (cycles/second) of the alternating current in an electrical power system, measured in hertz (Hz). In the United States, the rate is 60 Hz. The BA and several automatic systems maintain the frequency within tight limits.

fuel neutral: The concept that RTOs and FERC should not make rules that favor one fuel over another. In contrast, states have the option to make such rules.

generation: The production of electric energy from other sources of energy. Usually expressed in megawatts.

greenwashing: A form of marketing spin in which green values and green marketing are deceptively used to persuade the public that an organization's products, aims, and policies are environmentally friendly. The term is usually used to describe a company whose environmental rhetoric is nobler than its practices.

IESO: Independent Electricity System Operator.

Independent Electricity System Operator (IESO): The Regional Transmission Organization for Ontario, Canada.

independent system operator (ISO): See Regional Transmission Organization (RTO). There are some differences between an RTO and an ISO, but they are not relevant to most discussions.

intermittent power resource: A generator whose output power and availability are intermittent and not subject to the control of the grid operator or the plant operator (for example: wind, solar, run-of-river hydro). A generator also may be considered an intermittent power resource because of special contractual obligations or because it operates within the distribution system at less than 5 MW.

ISO: Independent system operator.

jump ball filing: When two entities (for example, ISO-NE and NEPOOL) are both concerned with a potential change of rules on the grid, and they file alternate proposals with FERC. Usually, these proposals are filed at the same time (as in "jump ball" in basketball), but sometimes there is an interval between the filings.

kludge: A quick-and-dirty solution that is clumsy, inelegant, inefficient, difficult to extend, and hard to maintain.

LMP: Locational Marginal Prices.

Load-serving entity (LSE): A company that buys electric energy, transmission service, and related services and resells it to end-use customers.

load shedding: Controlled or scheduled power outages (controlled blackouts) to balance the demand for electricity with a limited supply of electricity.

locational marginal prices (LMP): Grid price of electric energy in different locations on the grid. LMP prices are usually expressed as $/MWh. Locational pricing is affected by the general energy price (such as the price at a central hub), line losses, and congestion pricing.

LSE: Load-serving entity.

merit order: See Economic Dispatch.

Midcontinent ISO (MISO): The Regional Transmission Organization whose territory covers much of the Midwest and some areas of the South.

Minimum Offer Price Rule (MOPR): In the ISO-NE Forward Capacity Market, a rule requiring generators to offer capacity at a price at or above a floor set for each type of resource. MOPR does not allow generators receiving out-of-market revenue to reflect the supportive revenue in their capacity-price offers. Other RTO markets have similar rules.

MISO: Midcontinent ISO.

MOPR: Minimum Offer Price Rule.

N minus 1: The loss of a generation or transmission facility having a large impact on the system.

N minus 2: The loss of a second generation or transmission facility that would have a large impact on the system, following a N-minus-1 contingency.

nameplate capacity: The maximum amount of electric power a piece of equipment is able to produce. This is expressed in kW or MW.

NEPOOL: New England Power Pool.

NERC: North American Electric Reliability Corporation.

New England Power Pool (NEPOOL): The organization formed to ensure a dependable supply of electricity in New England. Various types of "stakeholders" are voting members of NEPOOL, in various types of subgroups (sectors). The six sectors are generation, supplier, publicly owned entities, alternative resources, transmission operators, and end users. NEPOOL must approve memberships in the ISO-NE Board of Directors, and NEPOOL can file requests to FERC on equal footing with ISO-NE (See Jump Ball).

New York ISO (NYISO): The Regional Transmission Organization whose territory covers much of New York State.

non-spinning reserve: Offline generation that is not synchronized to the system but can be quickly electrically synchronized to the system to serve demand. See also fast-start resource and spinning reserve.

North American Electric Reliability Corporation (NERC): A not-for-profit international regulatory authority whose mission is to reduce risks to the reliability and security of the grid. NERC develops and enforces reliability standards. It also educates, trains, and certifies industry personnel. Its coverage includes the United States, Canada, and part of Mexico.

NYISO: New York ISO.

ohm: A measure of electrical resistance.

out-of-market compensation: Payments received by generators that do not go through the market bidding process. These can be payments for Reliability Agreements, Renewable Energy Credits, or other payments not directly related to buying and selling energy, capacity, or ancillary services.

out-of-market revenue: See out-of-market compensation.

peak-load generating unit: A generating unit that is used to meet system requirements during peak-load periods, when the demand on the system is the greatest. These units typically operate at a relatively high cost and run when the price of electric energy is high. Sometimes called "peaker" plants. See also baseload generating unit.

PJM: PJM is the Regional Transmission Organization whose territory includes many East Coast and Midwestern states such as Pennsylvania, New Jersey, and Maryland.

price maker: See Clearing Price.

price taker: See Clearing Price.

price separation: The difference in clearing prices at different locations. In an energy market, price separation is due to transmission constraints and congestion. See locational marginal prices.

Public Service Board (PSB): A state-level organization that regulates in-state utilities. Their regulatory mandate usually includes electric, water, and telecomm utilities. Public Utility Commissions are very similar to Public Service Boards: the name, method of appointment of the membership, and board responsibilities vary by state.

PSB: Public Service Board.

Production Tax Credit (PTC): The federal renewable electricity Production Tax Credit is an inflation-adjusted per-kilowatt-hour (kWh) credit against taxes. For example, a wind-turbine owner may sell a kilowatt-hour of electricity to a distribution utility. The wind-turbine owner would then receive a production tax credit from the federal government. The tax credit varies from year to year, according to changes in legislation, type of generating unit, and inflation adjustments. (As an order of magnitude, the credit is about 2 cents per kWh.)

Public Utilities Commission (PUC): See Public Service Board.

PUC: Public Utilities Commission.

Real-Time Energy Market: An operation that balances the difference between the day-ahead scheduled amounts of electric energy and the real-time load requirements. See Day-Ahead Energy Market.

REC: Renewable Energy Credit.

regional transmission organization (RTO): An organization to manage the generation, transmission, and distribution of electric energy in a region, often including several states. The RTO is the Balancing Authority for its region. In RTO areas, generators and distribution companies are usually separate entities. An independent system operator is a variety of RTO.

Regulated areas: Areas without RTOs. In these areas, vertically integrated utilities usually own both the generating and distribution facilities, and often co-own the transmission facilities. Their main regulators are usually state level: a Public Service Board or Public Utilities Commission.

Reliability agreement: Formerly known as RMR.

Reliability Must Run (RMR): An agreement made between the RTO and a generator owner requiring that the generator continue to operate, even when it is not economical to do so. The agreement assures system reliability and that the generation owner recovers the costs of operation.

Renewable Energy Certificate: Renewable Energy Credit.

Renewable Energy Credit (REC): A certificate which represents actual electricity generation from a grid-connected, renewable source. Renewable sources are defined by the state. These certificates can be sold and traded. Their value can be set by a REC market or by state regulations.

Renewable Portfolio Standard (RPS): A state policy target requiring that distribution utilities meet the demand for electric energy by utilizing a certain percentage of renewable resources. Utilities can satisfy their RPS obligations by obtaining electric power from renewable generators in their own area or nearby localities, or by acquiring Renewable Energy Certificates from eligible renewable resources. In some states, RECs can be purchased from distant states.

Restructured states or areas: States (or areas within states) that have joined an RTO market.

RMR: Reliability Must Run.

Rolling blackouts: A form of load shedding, in which a utility schedules power outages for one area and then returns the power to that area and schedules a power outage for another area. The blackout (load shedding) "rolls" from place to place. See Load shedding.

RTO: Regional Transmission Organization.

Southwest Power Pool (SPP): The Regional Transmission Organization whose territory covers much of the Midwest and some areas of the South. SPP territories lie mostly to the west of MISO territories.

spinning reserve: Operating reserve which is synchronized to the grid system and can supply its full energy to the grid within ten minutes of receiving a request to come online.

SPP: Southwest Power Pool.

synchronous condenser: An electric motor synchronized to the transmission system. The motor shaft is spinning freely. Synchronous condensers adjust the VARs on the system.

VAR: Volt-ampere reactive.

vertically integrated utility: A utility, generally not in an RTO area, which owns both generation and distribution facilities. These utilities are usually basically responsible for all aspects of customer service. Their main regulators are state Public Service Boards.

volt-ampere reactive (VAR): Component of electric power that contributes no useful work but must be kept in balance to keep grid power in balance with itself.

volt: Unit that describes electromotive force. A flashlight battery might be 3 volts, while a transmission line might be 110,000 volts (110 Kv).

ZEC: Zero Emission Credit.

Zero Emission Credit (ZEC): Payments that electricity generators receive to compensate them for not emitting greenhouse gases in the production of electricity. Usually used in terms of nuclear generators. Defined at the state level, and usually parallel in structure and use to Renewable Energy Credits.

ENDNOTES

[1] Michael Lewis, *The Big Short* (W. W. Norton & Company, 2011).

[2] Meredith and George Angwin, *Voices for Vermont Yankee* (Carnot Communications, 2013).

[3] Meredith Angwin, *Campaigning for Clean Air: Strategies for Pro-Nuclear Advocacy*, (Carnot Communications, 2016).

[4] Elijah R. Perry, Meredith J. Angwin, Mario Rabinowitz, John F. Shimschock. Means for protecting underground electrical equipment from thermal runaway. US Patent 4,097,682, filed August 27, 1976, and issued June 27, 1978. *https://patents.google.com/patent/US4097682/un.*

[5] Meredith J. Angwin, *Yes Vermont Yankee, http://yesvy.blogspot.com/.*

[6] The newspaper article was based on the ISO-NE press release "Final Capacity Auction Results: Surplus Resources Available for 2013–2014," *Business Wire*, August 30, 2010. *https://www.businesswire.com/news/home/20100830006350/en/Final-Capacity-Auction-Results-Surplus-Resources-2013–2014.*

[7] This graphic is updated every few minutes. This snapshot was taken from the home page of the ISO-New England website at 2:45 pm on April 23, 2016. *https://www.iso-ne.com.*

[8] Basic review of the Otter Tail case in "Otter Tail Power Co. v. United States," *Wikipedia*, updated May 7, 2019, *https://en.wikipedia.org/wiki/Otter_Tail_Power_Co._v._United_States.*

[9] For discussion see: Lincoln L. Davies, Alexandra B. Klass, Hari M. Osofsky, Joseph B. Tomain, and Elizabeth J. Wilson, *Energy Law and Policy*, 2nd ed., Academic Casebook Series, (West Academic Publishing 2018), 143-9.

[10] Figure taken from the web page "Confronting the Duck Curve: How to Address Over-Generation of Solar Energy," Office of Energy Efficiency &

Renewable Energy, Department of Energy, *https://www.energy.gov/eere/articles/confronting-duck-curve-how-address-over-generation-solar-energy.*

[11] Andrew Stein, "Grid Operator tells Shumlin state knows why wind energy cutoff is required." *VTDigger*, August 7, 2013 *https://vtdigger.org/2013/08/07/grid-operator-tells-shumlin-state-knows-why-wind-energy-cutoff-is-required/*

[12] Meredith Angwin, "Wind on the Grid: Location, Location, Location," *Yes Vermont Yankee*, October 7, 2013, *http://yesvy.blogspot.com/2013/10/wind-on-grid-location-location-location.html#.XbidnafMxTa.*

[13] "Regional Transmission Organizations (RTO)/Independent System Operators (ISO)," Federal Energy Regulatory Commission, October 17, 2019, *https://www.ferc.gov/industries/electric/indus-act/rto.asp.*

[14] "Regional transmission organization (North America)," *Wikipedia*, last updated October 23, 2019, *https://en.wikipedia.org/wiki/Regional_transmission_organization_(North_America).*

[15] "State-by-State Information (web page)," American Coalition of Competitive Energy Suppliers (ACCES), undated, *http://competitiveenergy.org/consumer-tools/state-by-state-links/.*

[16] Vamsi Chadalavada, "Cold Weather Operations: December 24, 2017 – January 8, 2018," ISO New England, January 16, 2018, 14, *https://www.iso-ne.com/static-assets/documents/2018/01/20180112_cold_weather_ops_npc.pdf.*

[17] Meredith Angwin, "Oil Kept the Power Grid Running in Recent Cold Snap," *Valley News*, January 27, 2018, *http://www.vnews.com/Column-OilSavesGrid-ma-14865099.*

[18] Chadalavada, "Cold Weather Operations."

[19] Chadalavada, "Cold Weather Operations," 22.

[20] Chadalavada, "Cold Weather Operations," 17.

[21] Jon Chesto, "Russian LNG is unloaded in Everett; the supplier (but not gas) faces US sanctions," *The Boston Globe* (website), January 29, 2018 8:15 p.m., *https://www.bostonglobe.com/business/2018/01/29/tanker-unloads-lng-everett-terminal-that-contains-russian-gas/rewj1wKjajaKtLp79irzTI/story.html.*

[22] Rod Adams, "Performance of the New England power grid during extreme cold Dec 25-Jan 8," *Atomic Insights* (blog), January 26, 2018, *https://atomicinsights.com/performance-new-england-power-grid-extreme-cold-dec-25-jan-8/.*

[23] "FERC accepts ISO-NE's proposed winter 2013/2014 reliability program," *ISO Newswire*, September 18, 2013, *http://isonewswire.com/updates/2013/9/18/ferc-accepts-iso-nes-proposed-winter-20132014-reliability-pr.html.*

[24] Gavin Bade, "Powelson: FERC 'will not destroy the marketplace' in DOE cost recovery rulemaking," *Utility Dive*, October 5, 2017, *https://www.utilitydive.com/news/powelson-ferc-will-not-destroy-the-marketplace-in-doe-cost-recovery-rule/506577/.*

[25] "FERC accepts ISO-NE's ... program," *ISO Newswire.*

[26] "Update on the 2017/2018 Winter Reliability Program," *ISO Newswire*, October 27, 2017, ***http://isonewswire.com/updates/2017/10/27/update-on-the-20172018-winter-reliability-program.html.***

[27] Severin Borenstein and James Bushnell, "The U.S. Electricity Industry after 20 Years of Restructuring," *Energy Institute at Haas* (website), revised May 2015, figure 6, ***https://haas.berkeley.edu/wp-content/uploads/WP252.pdf.*** The authors request the following form of citation: "Borenstein, S, Bushnell, JB. The U.S. Electricity Industry after 20 Years of Restructuring. *Annu. Rev. Econ.* 7: Submitted. Doi: 10.1146/annureveconomics-080614-115630."

[28] "History of FERC," *Students Corner*, Federal Energy Regulatory Commission, accessed October 30, 2019, ***https://www.ferc.gov/students/ferc/history.asp.***

[29] Raymond L. Gifford and Matthew S. Larson, "For RTOs & ISOs: 'Don't call it a market' (props to LL Cool J)," *Utility Dive,* November 12, 2018, ***https://www.utilitydive.com/news/for-rtos-isos-dont-call-it-a-market-props-to-ll-cool-j/541895/.***

[30] Bill Julian, "Don't let ghosts of Enron haunt us again," *The Sacramento Bee*, September 24, 2016, ***https://www.sacbee.com/opinion/op-ed/article103535032.html.***

[31] Travis Kavulla, "There Is No Free Market for Electricity: Can There Ever Be?" *American Affairs* Volume I, Number 2 (Summer 2017), ***https://americanaffairsjournal.org/2017/05/no-free-market-electricity-can-ever/.***

[32] Borenstein and Bushnell, "The U.S. Electricity Industry."

[33] William B. Marcus, "Does Deregulation Raise Electric Rates? A Cross Sectional Analysis," JBS Energy, Inc., December 2011, ***http://citeseerx.ist.psu.edu/viewdoc/download?doi=10.1.1.307.9970&rep=rep1&type=pdf.*** JBS Energy has been dissolved; Mr. Marcus is now with MCPM Economics.

[34] David Gattie, "Residential Rates: Regulated vs. Deregulated Markets," *Energy: In The World As It Is* (blog), July 27, 2018, ***https://davidgattie blog.wordpress.com/2018/07/27/residential-rates-regulated-vs-deregulated-markets/.***

[35] *Texas Coalition for Affordable Power* (website), ***https://tcaptx.com.***

[36] R.A. "Jake" Dyer, Electricity Prices in Texas: A Snapshot Report, 2019 Edition," *Texas Coalition for Affordable Power* (website), May 2019, ***https://tcaptx.com/reports/snapshot-report-electricity-prices-texas-may-2019.***

[37] Keith Poli, "ERCOT's Final Summer 2019 Assessment & Report Findings: Declining Reserve Margins and Limited New Gas Generation," *Constellation: An Exelon Company* (blog), May 23, 2019, ***https://blogs.constellation.com/energy-management/ercots-preliminary-summer-2019-assessment-a-declining-reserve-margin/.***

[38] FERC, *Order Accepting Tariff Revisions, ISO New England, Docket ER19-1428-003,* Issued June 18, 2020 ***https://www.iso-ne.com/static-assets/documents/2020/06/er19-1483-003_6-18-20_order_accept_iep.pdf***

[39] Michael Brooks, "Abundance of Summer Capacity — Except in Texas," *RTO Insider*, May 16, 2019, *https://rtoinsider.com/abundance-summer-capacity-except-texas-117027/*.

[40] Susan Tierney, Todd Schatzki, and Rana Mukerji, "Uniform-Pricing versus Pay-as-Bid in Wholesale Electricity Markets: Does it Make a Difference?" *Analysis Group* for NYISO, 2008, *https://kylewoodward.com/blog-data/pdfs/references/tierney+schatzki+mukerji-new-york-iso-2008A.pdf*.

[41] Sarah Shemkus, "Critics say lifeline for Massachusetts LNG power plant sets bad precedent," *The Energy News Network* (a website project of Fresh Energy), January 28, 2019, *https://energynews.us/2019/01/28/northeast/critics-say-lifeline-for-massachusetts-lng-power-plant-sets-bad-precedent/*.

[42] Meredith Angwin, "Pay for Performance on the U.S. Grid: No help to nuclear," *Yes Vermont Yankee*, January 4, 2017, *https://yesvy.blogspot.com/2017/01/pay-for-performance-on-us-grid-no-help.html#.Xbw-XafMxTZ*.

[43] Meredith Angwin, "'Pay for Performance' and the US grid," *Nuclear Engineering International*, February 3, 2016, *https://www.neimagazine.com/features/feature-pay-for-performance-and-the-us-grid-4800656/*. The NEI chart is based on one presented at an Entergy press conference in October 2015. Here is the accompanying press release: "Entergy to Close Pilgrim Nuclear Power Station in Massachusetts No Later than June 1, 2019," *Pilgrim Nuclear Power Station* (website), October 13, 2015, *http://www.pilgrimpower.com/entergy-to-close-pilgrim-nuclear-power-station-in-massachusetts-no-later-than-june-1-2019/*.

[44] Anne George, "ISO New England Update: Consumer Liaison Group Meeting," ISO New England, December 6, 2018, *https://www.iso-ne.com/static-assets/documents/2018/12/clg_meeting_george_iso_update_presentation_december_6_2018_final.pdf*. The chart originally appeared in ISO-NE's presentation to the Consumer Liaison Group in December 2017 and was repeated in their December 2018 presentation.

[45] "Our History," *ISO New England*, *https://www.iso-ne.com/about/what-we-do/history/*.

[46] "Annual Report 2018," New England Power Pool, November 30, 2018, *http://nepool.com/uploads/Annual_Report_2018.pdf*.

[47] David T. Doot, "New England Power Pool: Second Restated NEPOOL Agreement," New England Power Pool, section 1.28A on sheet 11A, effective October 1, 2019, *http://nepool.com/uploads/Op-2d_RNA.pdf*.

[48] Doot, "New England Power Pool."

[49] "NEPOOL Participants by Sector with Related Person," New England Power Pool, October 1, 2019, 13-14, *http://nepool.com/uploads/C-Sector_Roster.pdf*.

[50] "Participants Committee," New England Power Pool, November 2019, *http://nepool.com/NPC_2019.php*.

[51] "NEPOOL Participants Committee Filing on Membership Amendments," New England Power Pool, August 13, 2018. Download file 12683546.PDF from *https://elibrary.ferc.gov/IDMWS/file_list.asp?document_id=14696812.*

[52] Rich Heidorn Jr., "FERC Rejects NEPOOL Press Membership Ban," *RTO Insider*, January 30, 2019, *https://rtoinsider.com/ferc-nepool-press-membership-ban-110091/.*

[53] Complaint of RTO Insider LLC against the New England Power Pool Participants Committee, August 31, 2018, download file 12693363.PDF from FERC Online eLibrary, *https://elibrary.ferc.gov/IDMWS/file_list.asp?document_id=14701573.*

[54] D. Maurice Kreis, "NEPOOL wins, transparency and electric ratepayers lose, at the FERC," *InDepthNH.org* (website of The New Hampshire Center for Public Interest Journalism), April 11, 2019, *http://indepthnh.org/2019/04/11/nepool-wins-transparency-and-electric-ratepayers-lose-at-the-ferc/.*

[55] "NEPOOL Officers 2019," *New England Power Pool* (website), *http://nepool.com/NEPOOL_Officers_2019.php.*

[56] "Board of Directors," *ISO New England* (website), *https://www.iso-ne.com/about/corporate-governance/board.*

[57] "Order on Proposed Tariff Revisions," Federal Energy Regulatory Commission, Docket No. ER15-2208-000, September 11, 2015, *https://www.ferc.gov/CalendarFiles/20150911153543-ER15-2208-000.pdf.*

[58] Gavin Bade, "PJM board sends competing capacity market reforms to FERC," *Utility Dive* (website), February 16, 2018, *https://www.utilitydive.com/news/pjm-board-sends-competing-capacity-market-reforms-to-ferc/517318/.*

[59] "ISO New England Inc. and New England Power Pool, Filings of Performance Incentives Market Rule Changes; Docket No. ER14-000," ISO New England, January 17, 2014, *https://www.iso-ne.com/static-assets/documents/regulatory/ferc/filings/2014/jan/er14_1050_000_1_17_14_pay_for_performace_part_1.pdf.* This FERC filing is 1746 pages long and not numbered sequentially. I refer to the entire filing as the "Joint Filing"; references within the Joint Filing are to page numbers within named documents that constitute the filing.

[60] In the Joint Filing, the description of the second settlement process begins on page 29 of Attachment I-1a and ends on page 76. Pages 77 and 78 contain a list of supporting documents and information.

[61] Joint Filing, Attachment I-1a, page 30.

[62] Joint Filing, Attachment N-1a, beginning on page 12.

[63] Joint Filing, Attachment N-1a, page 7.

[64] "Order on Tariff Filing and Instituting Section 206 Proceedings," Federal Energy Regulatory Commission, Docket Nos. ER14-1050-000, ER14-1050-001, and EL14-52-000, May 30, 2014, *https://www.iso-ne.com/static-assets/documents/regulatory/ferc/orders/2014/may/er14_1050_000_5_30_14_pay_for_performance_order.pdf.*

[65] Gavin Bade, "Unexpected outages, intense heat behind ISO-NE Labor Day price spike," *Utility Dive*, September 6, 2018, *https://www.utilitydive.com/news/unexpected-outages-intense-heat-behind-iso-ne-labor-day-price-spike/531751/.*

[66] Vamsi Chadalavada, "September 3 OP-4 Event and Capacity Scarcity Condition," ISO New England, September 12, 2018, 27, *https://www.iso-ne.com/static-assets/documents/2018/09/september-2018-op4-coo-report.pdf.*

[67] George Katsigiannakis, Shanthi Muthiah, Himanshu Pande, Rachel Green, and Josh Ghosh, "How ISO-NE's Pay-for-Performance Initiative Will Shake Up New England," ICF International, 2014, *http://www.ourenergypolicy.org/wp-content/uploads/2014/11/ISO_NE_Pay_for_Performance_Initiative.pdf.*

[68] Meredith Angwin, "Pay for Performance on the U.S. Grid: No help to nuclear," *Yes Vermont Yankee*, January 4, 2017, *https://yesvy.blogspot.com/2017/01/pay-for-performance-on-us-grid-no-help.html#.Xbw-XafMxTZ.*

[69] "Operational Fuel-Security Analysis," ISO New England, January 17, 2018, *https://iso-ne.com/static-assets/documents/2018/01/20180117_operational_fuel-security_analysis.pdf.*

[70] "Exelon Generation Files to Retire Mystic Generating Station in 2022, Absent Any Regulatory Solution," *Exelon* (website), March 29, 2018, *https://www.exeloncorp.com/newsroom/exelon-generation-files-to-retire-mystic-generating-station-in-2022.*

[71] Meredith Angwin, "Column: Rolling Blackouts Coming to a Power Grid Near You," *Valley News* (E-Edition), May 19, 2018, *https://www.vnews.com/Column-Rolling-Blackouts-Possible-for-New-England-17599767.*

[72] D. Maurice ("Don") Kreis, "ISO New England's Study Was Flawed," *Valley News*, Letters-to-the-editor, May 23, 2018, *https://www.vnews.com/Forum-May-24-17711851.*

[73] Paul Peterson, Doug Hurley, and Pat Knight, "Understanding ISO New England's Operational Fuel Security Analysis," Synapse Energy Economics, Inc., May 3, 2018, *https://www.clf.org/wp-content/uploads/2018/05/Understanding-ISO-NE-OFSA1.pdf.* This document is referred to in the text as "the Synapse Report."

[74] "Addendum to ISO Operational Fuel-Security Analysis (OFSA)," ISO New England, April 26, 2018, *https://www.iso-ne.com/static-assets/documents/2018/04/addendum-to-iso-operational-fuel-security-analysis.pdf.*

[75] RENEW Northeast *http://renew-ne.org*

[76] Meredith Angwin, "Cold Weather Winners and Losers on the Vermont Grid," *Yes Vermont Yankee*, January 24, 2013, *https://yesvy.blogspot.com/2013/01/cold-weather-winners-and-losers-on.html#.Xwt6di3MzUJ.*

[77] Mary C. Serreze, "Massachusetts regulators approve Berkshire Gas 5-year plan, while insisting company must work to lift moratorium," *MassLive.com*

(website), January 7, 2019, *https://www.masslive.com/business-news/2017/07/utility_regulators_berkshire_gas_need_no.html.*

78 "Operational Fuel-Security Analysis," ISO-NE, January 17, 2018, 23.

79 "Operational Fuel-Security Analysis," ISO-NE, January 17, 2018, 23.

80 The Synapse Report, page 1.

81 "Order Accepting and Suspending Filing and Establishing Hearing Procedures," Federal Energy Regulatory Commission, July 13, 2018, *https://www.ferc.gov/CalendarFiles/20180713175746-ER18-1639-000.pdf.*

82 "Energy Security Improvements," ISO New England, April 2019 — Version 1, *https://www.iso-ne.com/static-assets/documents/2019/04/a00_iso_discussion_paper_energy_security_improvements.pdf.*

83 Michael Kuser, "ISO-NE Filing, Whitepaper Address Energy Security," *RTO Insider*, April 4, 2019, *https://rtoinsider.com/iso-ne-whitepaper-energy-security-113956/.*

84 Chadalavada, "Cold Weather Operations," 17.

85 Michael Kuser, "NEPOOL MC Debates Energy Security Models," *RTO Insider*, June 18, 2019, *https://rtoinsider.com/nepool-debates-energy-security-models-138484/.*

86 "Energy Security Improvements," ISO New England, April 2019 — Version 1, page 20 and following *https://www.iso-ne.com/static-assets/documents/2019/04/a00_iso_discussion_paper_energy_security_improvements.pdf*

87 Kuser, "ISO-NE Filing."

88 "DEC Lunch: Assessing and mitigating the risk of cascading blackouts," *The Arthur L. Irving Institute for Energy and Society* (website), March 7, 2019, *https://irving.dartmouth.edu/events/event?event=54483.*

89 Margaret J. Eppstein and Paul D. H. Hines, "A 'Random Chemistry' Algorithm for Identifying Collections of Multiple Contingencies That Initiate Cascading Failure," *IEEE Transactions on Power Systems*, Vol. 27, No. 3, August 2012, *http://www.cs.uvm.edu/~meppstei/personal/IEEEPES2012.pdf.*

90 Scott DiSavino, "New England grid can function without Vermont Yankee reactor," Reuters News Agency, December 10, 2012, *https://www.reuters.com/article/utilities-entergy-vermontyankee/new-england-grid-can-function-without-vermont-yankee-reactor-idUSL1E8NA50O20121210.*

91 David Brooks, "Northern Pass died — but the debate didn't," *Concord Monitor*, December 28, 2018, *https://www.concordmonitor.com/story-of-year-energy-northern-pass-22403247.*

92 Mike Polhamus, "TDI New England approved for $1.2 billion transmission project," *VTDigger*, January 6, 2016, *https://vtdigger.org/2016/01/06/tdi-new-england-approved-for-1-2-billion-transmission-project/.*

[93] "Order No. 1000 — Transmission Planning and Cost Allocation," Federal Energy Regulatory Commission, updated August 22, 2019, *https://www.ferc.gov/industries/electric/indus-act/trans-plan.asp.*

[94] "Electricity Transmission Cost Allocation," *EveryCRSReport.com* (website), December 18, 2012, *https://www.everycrsreport.com/reports/R41193.html.*

[95] "Order No. 1000 ..." Federal Energy Regulatory Commission.

[96] Federal Register, Volume 76, Number 155, August 11, 2011, *https://www.ferc.gov/industries/electric/indus-act/trans-plan/fr-notice.pdf.*

[97] "Our Work," *VELCO* (website of Vermont Electric Power Company), undated, *https://www.velco.com/our-work.*

[98] "Final Minutes," VELCO Operating Committee, October 17, 2013, *https://opcom.velco.com/library/document/download/1335/Final%252520Minutes%25252010-17-13.pdf.*

[99] "Tragedy of the commons," *Wikipedia*, last update November 10, 2019, *https://en.wikipedia.org/wiki/Tragedy_of_the_commons.*

[100] Jason Marshall, "Order 1000 in New England," New England States Committee on Electricity, September 7, 2017, *https://www.iso-ne.com/static-assets/documents/2017/09/clg_meeting_marshall_panelist_presentation_september_7_2017_final.pdf.*

[101] "About NESCOE (web page)," NESCOE: New England States Committee on Electricity, undated, *http://nescoe.com/about-nescoe/.*

[102] Herman K. Trabish, "Has FERC's landmark transmission planning effort made transmission building harder?" *Utility Dive*, July 17, 2018, *https://www.utilitydive.com/news/has-fercs-landmark-transmission-planning-effort-made-transmission-building/527807/.*

[103] Marshall, "Order 1000 in New England."

[104] Tony Clark, Competitive Transmission Development Technical Conference, FERC Docket No. AD16-18-000, June 27, 2016.

[105] Anne George, "ISO New England Update; Consumer Liaison Group Meeting," ISO New England, June 20, 2019, *https://www.iso-ne.com/static-assets/documents/2019/06/clg_meeting_george_iso_update_presentation_june_20_2019_final.pdf.*

[106] "2018 Report of the Consumer Liaison Group," ISO New England, March 12, 2019, *https://www.iso-ne.com/static-assets/documents/2019/03/2018_report_of_the_consumer_liaison_group_final.pdf.*

[107] Press Release by Green Mountain Power: "Green Mountain Power uses Tesla Power Walls to reduce electricity demand," *VTDigger*, August 14, 2018, *https://vtdigger.org/2018/08/14/green-mountain-power-uses-tesla-power-walls-reduce-electricity-demand/.*

[108] The Texas RTO, ERCOT, does have a grid that is stressed in summer. They do not have high reserve margins. This is described in the section on "The capacity auction" in chapter 14. One difficulty about writing about the grid is that, with so many jurisdictions involved, there is probably somewhere that has a different version of the problem.

[109] Gavin Bade, "Unexpected outages, intense heat behind ISO-NE Labor Day price spike," *Utility Dive*, September 6, 2018, *https://www.utilitydive.com/news/unexpected-outages-intense-heat-behind-iso-ne-labor-day-price-spike/531751/.*

[110] Chadalavada, "September 3 OP-4 Event."

[111] Meredith Angwin, "The Not-Stressed Grid in Summer," *Yes Vermont Yankee*, July 5, 2018, *https://yesvy.blogspot.com/2018/07/the-not-stressed-grid-in-summer.html#.XcquRafMxTY.*

[112] "Vermont communities open 'cooling stations,'" *vermontbiz* (website), July 2, 2018, *https://vermontbiz.com/news/2018/july/02/vermont-communities-open-cooling-stations?utm_source=VBM+Mailing+List&utm_campaign=e5bc2355ce-ENEWS_2018_07_02.*

[113] Dan D'Ambrosio, "GMP uses battery power to save $200k," *Burlington Free Press*, October 27, 2016, *https://www.burlingtonfreepress.com/story/news/2016/10/27/green-mountain-power-saves-200k-battery-power/92824712/.*

[114] "Stored energy helped GMP save $500K during heatwave," *vermont-biz* (website), July 12, 2018, *https://vermontbiz.com/news/2018/july/12/stored-energy-helped-gmp-save-500k-during-heatwave.*

[115] Press Release by Green Mountain Power: "Green Mountain Power uses Tesla Power Walls to reduce electricity demand." *VTDigger,* August 14, 2018, *https://vtdigger.org/2018/08/14/green-mountain-power-uses-tesla-power-walls-reduce-electricity-demand/*

[116] Press Release by Green Mountain Power: "Green Mountain Power turns to stored energy in heat wave," *VTDigger*, July 9, 2018, *https://vtdigger.org/2018/07/09/green-mountain-power-turns-stored-energy-heat-wave/.*

[117] Bloomberg, "PG&E Officially Files for Bankruptcy Under the Financial Strain of California Wildfires," *Fortune*, January 29, 2019, *https://fortune.com/2019/01/29/pge-bankruptcy-filing-wildfire-bill/.*

[118] Chadalavada, "September 3 OP-4 Event."

[119] This snapshot was taken from the home page of the ISO New England website (*https://www.iso-ne.com*), which is updated every few minutes. It is a summary of actual demand on January 14, 2019.

[120] Hat tip to Warren Van Wyck, who suggested this clear comparison for "what is baseload."

[121] "Real-Time Maps and Charts," *ISO New England* (website), snapshot of charts taken at 2:51 pm on July 8, 2019, *https://www.iso-ne.com/isoexpress/.*

[122] Chadalavada, "Cold Weather Operations."

[123] "Table 4.08.B. Capacity Factors for Utility Scale Generators Primarily Using Non-Fossil Fuels" Electric Power Annual published by the U.S. Energy Information Agency, October 18, 2019, *https://www.eia.gov/electricity/annual/html/epa_04_08_b.html.*

[124] "Chief Joseph Dam," *Wikipedia*, updated October 15, 2019, *https://en.wikipedia.org/wiki/Chief_Joseph_Dam.*

[125] Mark Z. Jacobson, Mark A. Delucchi, Mary A. Cameron, and Bethany A. Frew, "Low-cost solution to the grid reliability problem with 100% penetration of intermittent wind, water, and solar for all purposes," *Proceedings of the National Academy of Sciences*, 112, no. 49 (December 8, 2015): 15060-15065, *https://www.pnas.org/content/112/49/15060.*

[126] Christopher T. M. Clack et al., "Evaluation of a proposal for reliable low-cost grid power with 100% wind, water, and solar," *Proceedings of the National Academy of Sciences* 114, no. 26 (June 27, 2017): 6722-2627, *https://www.pnas.org/content/114/26/6722.*

[127] Supporting information for the above article by Clack et al., *https://www.pnas.org/content/pnas/suppl/2017/06/16/1610381114.DCSupplemental/pnas.1610381114.sapp.pdf.*

[128] Mark Z. Jacobson, Mark A. Delucchi, Mary A. Cameron, and Bethany A. Frew, "The United States can keep the grid stable at low cost with 100% clean, renewable energy in all sectors despite inaccurate claims" (letter), *Proceedings of the National Academy of Sciences* 114, no. 26 (June 27, 2017), *https://www.pnas.org/content/114/26/E5021.*

[129] Julian Spector, "Mark Jacobson Drops Lawsuit Against Critics of His 100% Renewables Plan," *gtm:* (website of Greentech Media), February 26, 2018, *https://www.greentechmedia.com/articles/read/mark-jacobson-drops-lawsuit-against-critics-of-his-100-renewables.*

[130] Besides my general knowledge of the grid and several visits to working dams, I also headed a project on predicting and preventing corrosion in the penstocks of several medium-size dams in mountainous country. This project was not published: it was only a report to the client, so I cannot provide a link. While I would not claim hydro power as an area of deep expertise for me, I have enough knowledge to be seriously skeptical about the idea of adding ten times as many turbines to existing hydro plants.

[131] Mike Conley and Tim Maloney, "Road Map to Nowhere: The Myth of Powering the Nation with Renewable Energy," *Road Map to Nowhere* (website), December 2017, *https://www.roadmaptonowhere.com.*

[132] Mathijs Beckers, "The non-solutions project," *CreateSpace Independent Publishing Platform* (January 18, 2017), *https://www.amazon.com/gp/product/B01N6SN5E1/ref=dbs_a_def_rwt_hsch_vapi_tkin_p1_i1.*

[133] "Legislation to Address the Urgent Threat of Climate Change," letter to Congress, January 10, 2019, *https://1bps6437gg8c169i0y1drtgz-wpengine.netdna-ssl.com/wp-content/uploads/2019/01/Progressive-Climate-Leg-Sign-On-Letter-2.pdf.*

[134] Matt Nesvisky, "A Role for Fossil Fuels in Renewable Energy Diffusion" (digest of NBER Working Paper No. 22454), The National Bureau of Economic Research, November 13, 2019, *https://www.nber.org/digest/oct16/w22454.html.*

[135] Chris Mooney, "Turns out wind and solar have a secret friend: Natural gas," *The Washington Post*, August 11, 2016, *https://www.washingtonpost.com/news/energy-environment/wp/2016/08/11/turns-out-wind-and-solar-have-a-secret-friend-natural-gas/.*

[136] Christine Hallquist, "The Distribution Grid," Presentation at Osher Grid class 3 May 2016, *https://www.youtube.com/watch?v=4zN0D_CAkKE*

[137] Hallquist presentation, at around the one hour twenty-five minute point of her talk.

[138] "Combined-cycle power plant," *Wikipedia*, updated November 13, 2019, *https://en.wikipedia.org/wiki/Combined_cycle_power_plant.*

[139] Euan Mearns, "CO_2 Emissions Variations in CCGTs Used to Balance Wind in Ireland," *Energy Matters* (blog), April 15, 2016, *http://euanmearns.com/co2-emissions-variations-in-ccgts-used-to-balance-wind-in-ireland/.*

[140] "Energy in Ireland: 1990-2016," Sustainable Energy Authority of Ireland, December 2017, *https://www.seai.ie/publications/Energy-in-Ireland-1990-2016-Full-report.pdf.*

[141] Dan E. Way, "Duke Energy application points finger at solar for increased pollution," *North State Journal*, August 14, 2019, *https://nsjonline.com/article/2019/08/duke-energy-application-points-finger-at-solar-for-increased-pollution/.*

[142] Kristi E. Swartz, "Can solar increase emissions? A debate erupts," E&E News, August 21, 2019, *https://www.eenews.net/stories/1061015535?fbclid=IwAR3Wz-1p2fyjtQDqihk5SAP_f2kotsiWD-tqusvah2AJ7kqLqVd9-eRNkcZY.*

[143] Joel Banner Baird, "Can wind power plug-and-play with the grid?" *Burlington Free Press*, February 2, 2014, *https://www.burlingtonfreepress.com/story/news/2014/02/02/can-wind-power-plug-and-play-with-the-grid-/5082641/.*

[144] Elizabeth Gribkoff, "Proposed solar fee raises questions about who pays for grid upgrades," *VTDigger*, June 2, 2019, *https://vtdigger.org/2019/06/02/proposed-solar-fee-raises-questions-pays-grid-upgrades/.*

[145] Donovan Alexander, "Tesla Battery Installed in South Australia Saved the Region $40 Million in Its First Year," *Interesting Engineering* (website), December 8, 2018, *https://interestingengineering.com/*

tesla-battery-installed-in-south-australia-saved-the-region-40-million-in-its-first-year.

[146] Kevin McCallum, "Charging Forward: Battery Power Projects Are Surging in Vermont," *Seven Days* (website), November 27, 2019, *https://www.sevendaysvt.com/vermont/charging-forward-battery-power-projects-are-surging-in-vermont/Content?oid=29025774.*

[147] Nestor Sepulveda, Jesse Jenkins, Fernando de Sisternes, and Richard Lester, "The Role of Firm Low-Carbon Electricity Resources in Deep Decarbonization of Power Generation," *Joule* 2, no. 11 (November 21, 2018): 2403-2420, *https://www.sciencedirect.com/science/article/pii/S2542435118303866.*

[148] "Reserve electric generating capacity helps keep the lights on," *eia* (website of U.S. Energy Information Administration), June 1, 2012, *https://www.eia.gov/todayinenergy/detail.php?id=6510.*

[149] "2019-2020 Winter Reliability Assessment." NERC, North American Electric Reliability Corporation, November 2019, *https://www.nerc.com/pa/RAPA/ra/Reliability%20Assessments%20DL/NERC%20WRA%202019_2020.pdf#search=winter%20reserve%20margins*

[150] News Release by the Electric Reliability Council of Texas: "New report shows tightening electricity reserve margins," *ercot* (website of ERCOT), December 4, 2018, *http://www.ercot.com/news/releases/show/168033.*

[151] Hewitt Crane, Edwin Kinderman, Ripudaman Malhotra, *A Cubic Mile of Oil: Realities and Options for Averting the Looming Global Energy Crisis* (Oxford University Press, 2010).

[152] Ripudaman Malhotra, "Replacing coal with wind and solar power," *A Cubic Mile of Oil* (blog), September 20, 2019, *https://cmo-ripu.blogspot.com/2019/09/replacing-coal-with-wind-and-solar-power.html.*

[153] Liam O'Brien, "Top 5 Issues with Nickel-Iron (NiFe) Batteries," DIY Homestead Projects, April 21, 2018, *https://www.diyhomesteadprojects.com/top-5-issues-with-nickel-iron-nife-batteries/*

[154] Nancy Pfotenhauer, "Big Wind's Bogus Subsidies," *U.S. News & World Report*, May 12, 2014, *https://www.usnews.com/opinion/blogs/nancy-pfotenhauer/2014/05/12/even-warren-buffet-admits-wind-energy-is-a-bad-investment.*

[155] Zach Starsia, "What Corporate Buyers of Renewable Energy Need to Know About Wind and Solar Tax Credits," *LevelTen Energy* (website), April 5, 2019, *https://leveltenenergy.com/blog/energy-procurement/renewable-energy-tax-credits/.*

[156] Molly F. Sherlock, "The Energy Credit: An Investment Tax Credit for Renewable Energy," Congressional Research Service, updated November 2, 2018, *https://crsreports.congress.gov/product/pdf/IF/IF10479.*

[157] "Tax Policy (web page)," *AWEA* (website of the American Wind Energy Association), undated, *https://www.awea.org/policy-and-issues/tax-policy.*

[158] Richard Bowers, "Tax credit phaseout encourages more wind power plants to be added by end of year," *eia* (website of U.S. Energy Information Administration), May 15, 2019, *https://www.awea.org/policy-and-issues/tax-policy.*

[159] "United States wind energy policy," *Wikipedia*, updated September 30, 2019, *https://en.wikipedia.org/wiki/United_States_wind_energy_policy #History_of_the_Production_Tax_Credit.*

[160] Press Release by Energize Vermont: "Wind developer Blittersdorf sues project neighbors," *VTDigger*, August 21, 2012, *https://vtdigger.org/ 2012/08/21/wind-developer-blittersdorf-sues-project-neighbors/.*

[161] "Georgia Mountain Community Wind December 2012" (newsletter), *http:// georgiamountainwind.com/wp-content/uploads/2013/05/GMCW-Newsletter-Dec-2012.pdf.*

[162] "U.S. solar sector launches lobbying push to preserve key subsidy," Reuters News Agency (website), July 17, 2019, *https://www.reuters.com/article/ us-usa-solar-idUSKCN1UC2IJ.* See how the credits get extended: "REV working with Welch to extend renewable energy tax credits," vermont-biz, November 15, 2019, *https://vermontbiz.com/news/2019/november/15/ rev-working-welch-extend-renewable-energy-tax-credits?utm_source=VBM+ Mailing+List&utm_campaign=a815d3902e-ENEWS_2019_11_15&utm_ medium=email&utm_term=0_b5e56b36a4-a815d3902e-286677853.*

[163] "Integrating Markets and Public Policy," *New England Power Pool* (website), undated, *http://nepool.com/IMAPP.php.*

[164] "The renewable energy credits market: where to buy RECs," *energysage* (website), updated June 7, 2019, *https://www.energysage.com/other-clean-options/ renewable-energy-credits-recs/where-to-buy-renewable-energy-credits-recs/.*

[165] "SREC Markets, Pennsylvania" *SRECTrade* (website of Superior Clean Energy Management), undated, *https://www.srectrade.com/markets/rps/srec/pennsylvania.*

[166] James D. Bride, New England CLG presentation, Energy Tariff Experts LLC, December 1, 2016, *https://www.iso-ne.com/static-assets/documents/ 2016/11/2016_12_01_clg_meeting_jim_bride_panelist_presentation.pdf.*

[167] Energy Tariff Experts LLC (website), *http://energytariffexperts.com.*

[168] Press Release by ISO New England: "New England's Wholesale Electricity Prices in 2017 Were the Second-Lowest Since 2003," *ISO New England* (website), March 6, 2018, *https://www.iso-ne.com/static-assets/documents/2018/03/20180306_pr_ 2017prices.pdf.*

[169] John Dillon, "Small Utilities Say Subsidized 'Net-Metering' Projects Could Trigger Rate Increases," *VPR (*website of Vermont Public Radio),

September 4, 2018, *https://www.vpr.org/post/small-utilities-say-subsidized-net-metering-projects-could-trigger-rate-increases#stream/0.*

[170] "FERC approves formula for BPA curtailment of wind generation in favor of excess hydro," *Hydro Review* (website), October 24, 2014, *https://www.hydroreview.com/2014/10/24/ferc-approves-formula-for-bpa-curtailment-of-wind-generation-in-favor-of-excess-hydro/.* Also see:

Amelia Templeton, "BPA Orders NW Wind Farms to Curtail Production," *OPB* (website of Oregon Public Broadcasting), April 30, 2012, *https://www.opb.org/news/article/bpa-orders-nw-wind-farms-to-curtail-production/?fbclid=IwAR39oWJeLSk-xvmbjxQh5zjG01-CoLp2CERnqqF4GvsETxIx7hTRgL6OX0E?fbclid=IwAR39oWJeLSk-xvmbjxQh5zjG01-CoLp2CERnqqF4GvsETxIx7hTRgL6OX0E.*

[171] Michael Greenstone and Ishan Nath, "Do Renewable Portfolio Standards Deliver?" Energy Policy Institute at the University of Chicago, April 22, 2019, *https://epic.uchicago.edu/research/do-renewable-portfolio-standards-deliver/.*

[172] Annette Smith: An examination of Vermont's energy policies, *VTDigger,* March 18,2018, *https://vtdigger.org/2018/03/18/annette-smith-examination-vermonts-energy-policies/*

[173] Leah Stokes, *Short Circuiting Policy: Interest Groups and the Battle Over Clean Energy and Climate Policy in the American States*, Oxford University Press, March 2020, *https://global.oup.com/academic/product/short-circuiting-policy-9780190074265?cc=us&lang=en&*. The quote is from the book announcement by Susan Liebell, UC Santa Barbara Department of Political Science, January 28, 2020, *https://www.polsci.ucsb.edu/news/announcement/990.*

[174] Annette Smith, "Understanding Vermont's Energy Policies," (website), *Vermonters for a Clean Environment* (website), March 2018, *http://www.vce.org/VCE_WhitePaper_UnderstandingVermontEnergyPolicies_12March2018.pdf.*

[175] Terri Hallenbeck, "Annette Smith Is a Lightning Rod in the Renewable-Energy Debate," *Seven Days*, February 10, 2016, *https://www.sevendaysvt.com/vermont/annette-smith-is-a-lightning-rod-in-the-renewable-energy-debate/Content?oid=3166359.*

[176] "Check Out Wind Power, Up Close (web page)," Green Mountain Power, May 15, 2019, *https://greenmountainpower.com/2019/05/15/check-out-wind-power-up-close/.*

[177] Press Release by Vermont Law School: "FTC warns Green Mountain Power to be clear about renewable energy claims," *VTDigger*, February 10, 2015, *https://vtdigger.org/2015/02/10/ftc-warns-green-mountain-power-clear-renewable-energy-claims/.*

[178] Herman K. Trabish, "NextEra drops Vermont RECs, adding weight to fraud claims," *Utility Dive*, May 21, 2014, *https://www.utilitydive.com/news/nextera-drops-vermont-recs-adding-weight-to-fraud-claims/265767/.*

[179] John Herrick, "Special Report: New renewable standard would revolutionize energy use in Vermont," *VTDigger*, February 17, 2015, *https://vtdigger.org/2015/02/17/special-report-new-renewable-standard-revolutionize-energy-use-vermont/.*

[180] "GMP's Energy Supply is 90% Carbon Free," Green Mountain Power, December 13, 2018, *https://greenmountainpower.com/2018/12/13/fuel-mix/.*

[181] Meredith Angwin, "What going to 90% renewable energy would do to Vermont's landscape," *Yes Vermont Yankee*, May 14, 2013, *https://yesvy.blogspot.com/2013/05/the-90-solution-what-90-renewables.html#.Xc_7Q6fMxTa.*

[182] Meredith Angwin, "The Measured Move to Renewables in Vermont," *Yes Vermont Yankee*, June 9, 2012, *https://yesvy.blogspot.com/2012/06/measured-move-to-renewables-valley-news.html#.XdKdFKfMxTb.*

[183] "Cow Power," *Green Mountain Power* (website), undated, *https://greenmountainpower.com/help/billing-payments/cow-power/.*

[184] Kristin Kelly (private communication), Green Mountain Power, November 12, 2019.

[185] "Service Area" (web page), Green Mountain Power, updated frequently, *https://greenmountainpower.com/map/service-area/.*

[186] "Air Quality (web page)," *NEI* (website of Nuclear Energy Institute), undated, *https://www.nei.org/advantages/air-quality.*

[187] Meredith Angwin, "The New York Clean Energy Standard," *Yes Vermont Yankee*, August 1, 2016, *https://yesvy.blogspot.com/2016/08/the-new-york-clean-energy-standard.html#.XdKh_afMxTZ.*

[188] Rod Adams, "Fighting climate change with best available tools," *Atomic Insights* (blog), August 1, 2016, *https://atomicinsights.com/fighting-climate-change-best-available-tools/.*

[189] Tim Knauss, "Dozens of CNY residents flood Albany meeting on nuclear subsidies," *syracuse.com*, August 1, 2016, updated January 4, 2019, *https://www.syracuse.com/news/2016/08/dozens_of_cny_residents_flood_albany_meeting_on_nuclear_subsidies.html.*

[190] Jeff St. John, "New York's Nuclear Zero-Carbon Credits Pass Federal Court Challenge," *gtm:* (website of Greentech Media), September 27, 2018, *https://www.greentechmedia.com/articles/read/new-yorks-nuclear-zero-carbon-credits-pass-federal-court-challenge#gs.8zkb0k.*

[191] Press Release by Environmental Defense Fund, "Appeals Court Upholds New York Clean Energy Plan," *EDF: Environmental Defense Fund* (website), September 27, 2018, *https://www.edf.org/media/appeals-court-upholds-new-york-clean-energy-plan.*

[192] Robert Bryce, "Indian Point nuclear-reactor shutdown a huge blow to New York's environment," *New York Post*, April 29, 2020, *https://nypost.com/2020/04/29/indian-point-nuclear-reactor-shutdown-a-huge-blow-to-nys-environment/.*

[193] Iulia Gheorghiu, "Connecticut moves to preserve Millstone Nuclear plant with 10-year power deal," *Utility Dive*, January 3, 2019, *https://www.utilitydive.com/news/connecticut-moves-to-preserve-millstone-nuclear-plant-with-10-year-power-de/545133/*. Also see: Catherine Morehouse, "State carbon-free policies increasingly inclusive of nuclear, but resource needs federal boost," *Utility Dive*, July 17, 2019, *https://www.utilitydive.com/news/state-carbon-free-policies-increasingly-inclusive-of-nuclear-but-resource/557535/*.

[194] Nadja Popovich, "How Does Your State Make Electricity?" *The New York Times*, December 24, 2018, *https://www.nytimes.com/interactive/2018/12/24/climate/how-electricity-generation-changed-in-your-state.html*.

[195] Meredith Angwin, "Clean Air versus Efficiency Charges. Clean Air Wins." *Yes Vermont Yankee*, August 22, 2016, *https://yesvy.blogspot.com/2016/08/clean-air-versus-efficiency-charges.html#.XdKq4afMxTZ*.

[196] "Governor Cuomo Announces Establishment of Clean Energy Standard that Mandates 50 Percent Renewables by 2030," *New York State* (website), August 1, 2016, *https://www.governor.ny.gov/news/governor-cuomo-announces-establishment-clean-energy-standard-mandates-50-percent-renewables*.

[197] Bride, New England CLG presentation.

[198] "Frequently Asked Questions," *eia* (website of U.S. Energy Information Administration), updated October 2, 2019, *https://www.eia.gov/tools/faqs/faq.php?id=97&t=3*.

[199] Table A.III.2 in "Technology-specific Cost and Performance Parameters," Intergovernmental Panel on Climate Change, 2014, *https://www.ipcc.ch/site/assets/uploads/2018/02/ipcc_wg3_ar5_annex-iii.pdf*. The official citation is:

> Schlömer S., T. Bruckner, L. Fulton, E. Hertwich, A. McKinnon, D. Perczyk, J. Roy, R. Schaeffer, R. Sims, P. Smith, and R. Wiser, 2014: Annex III: Technology-specific cost and performance parameters. In: *Climate Change 2014: Mitigation of Climate Change. Contribution of Working Group III to the Fifth Assessment Report of the Intergovernmental Panel on Climate Change* [Edenhofer, O., R. Pichs-Madruga, Y. Sokona, E. Farahani, S. Kadner, K. Seyboth, A. Adler, I. Baum, S. Brunner, P. Eickemeier, B. Kriemann, J. Savolainen, S. Schlömer, C. von Stechow, T. Zwickel and J.C. Minx (eds.)]. Cambridge University Press, Cambridge, United Kingdom and New York, NY, USA.

[200] Mike Twomey, "The Replacement for Vermont Yankee Was ... Natural Gas," *Yes Vermont Yankee*, January 7, 2016, *https://yesvy.blogspot.com/2016/01/the-replacement-for-vermont-yankee.html#.Xdkqs6fMxTb*.

[201] After Vermont Yankee shut down, let's look at the amount of carbon dioxide that was added to the air because gas turbines (assumed to be combined-cycle) replaced 4.9 million MWh per year of nuclear on the grid.

> 4.9 million MWh is 4.9×10^6 MWh. 1 MWh is 10^3 kWh, so we have 4.9×10^9 kWh.

According to IPCC, a natural gas combined-cycle plant emits 490 g CO_2 per kWh, while nuclear emits 12 g CO_2 per MWh. The difference is 480 grams (4.8×10^2) more CO_2 per kWh by using natural gas.

Multiplying 4.8×10^2 grams CO_2 per kWh by the 4.9×10^9 kWh of nuclear replaced by natural gas gives us an extra 2.4×10^{12} grams of CO_2 per year.

To get from grams to metric tons, we have to divide by 10^3 for grams per kilogram and by 10^3 again for kilograms per metric ton. We end up with 2.4×10^6 metric tons.

In conclusion, at least 2,400,000 more metric tons of carbon dioxide were emitted each year on the New England grid because Vermont Yankee closed.

202 "Single-cycle combustion turbine," *Wikipedia*, updated February 24, 2017, *https://en.wikipedia.org/wiki/Simple_cycle_combustion_turbine.*

203 "Combined-cycle power plant," *Wikipedia*, updated November 22, 2019, *https://en.wikipedia.org/wiki/Combined_cycle_power_plant#Efficiency_of_CCGT_plants.*

204 Meredith Angwin, "The Future of Nuclear in RTO Areas," *Yes Vermont Yankee*, November 21, 2016, *https://yesvy.blogspot.com/2016/11/the-future-of-nuclear-in-rto-areas.html#.XdWpoafMxTZ.*

205 Jacob Mays, David Morton, and Richard P. O'Neill, "Asymmetric Risk and Fuel Neutrality in Capacity Markets," United States Association for Energy Economics, February 8, 2019, *https://papers.ssrn.com/sol3/papers.cfm?abstract_id=3330932.*

206 Patricio Rocha-Garrido, "Wind Effective Load Carrying Capability (ELCC) Analysis," *PJM* (website), September 13, 2018, *https://www.pjm.com/-/media/committees-groups/committees/pc/20180913/20180913-item-05b-wind-effective-load-carrying-capability-elcc.ashx.*

207 "Table 6.07.B. Capacity Factors for Utility Scale Generators Primarily Using Non-Fossil Fuels," *eia* (website of U.S. Energy Information Administration), released October 24, 2019, *https://www.eia.gov/electricity/monthly/epm_table_grapher.php?t=epmt_6_07_b.*

208 "Order on Tariff Filing by ISO New England Inc. Docket No. ER18-619-000," Federal Energy Regulatory Commission, March 9, 2018. Download FERC Generated PDF 12586062.PDF from *https://elibrary.ferc.gov/idmws/file_list.asp?accession_num=20180309-4003.*

209 Gavin Bade, "Chatterjee opposes MOPR as 'standard solution' for state policies," *Utility Dive*, April 19, 2018, *https://www.utilitydive.com/news/chatterjee-opposes-mopr-as-standard-solution-for-state-policies/521731/.*

210 Sonal Patel, "PJM Will Hold Capacity Auction Under Current Rules in August," *POWER* (website), April 10, 2019, *https://www.powermag.com/pjm-will-hold-capacity-auction-under-current-rules-in-august/.*

[211] Iulia Gheorghiu, "FERC orders PJM to postpone capacity auction," *Utility Dive*, July 26, 2019, *https://www.utilitydive.com/news/ferc-orders-pjm-to-postpone-capacity-auction/559610/*.

[212] Anne George, "ISO New England Update; Consumer Liaison Group Meeting," ISO New England, March 14, 2019, *https://www.iso-ne.com/static-assets/documents/2019/03/clg_meeting_george_iso_update_presentation_march_14_2019_final.pdf*. The chart is slide 21.

[213] George, ISO New England Update, June 20, 2019, slide 16.

[214] "Order on Tariff Filing ... Docket No. ER18-619-000."

[215] Josh Siegel, "Republican FERC commissioner Robert Powelson to resign amid fight over coal bailout," *Washington Examiner*, June 28, 2018, *https://www.washingtonexaminer.com/policy/energy/republican-ferc-commissioner-robert-powelson-to-resign-amid-fight-over-coal-bailout*.

[216] Robert F. Powelson, "The ISO-NE Competitive Auctions with Sponsored Policy Resources Proposal," Federal Energy Regulatory Commission, March 9, 2018, *https://ferc.gov/media/statements-speeches/powelson/2018/03-09-18-powelson.asp#.XdqBHafMxTZ*.

[217] Iulia Gheorghiu, "Capacity pricing changes: How each power market plans to account for resource adequacy," *Utility Dive*, December 18, 2018, *https://www.utilitydive.com/news/capacity-pricing-changes-how-each-power-market-plans-to-account-for-resour/542449/?fbclid=IwAR1PoNJucoIqp8nzifLxLat--QUTfdN_DHh6BWmJmGTFPwBRiwe5HVpdIe0*.

[218] Coley Girouard, "Understanding IRPs: How Utilities Plan for the Future," *Advanced Energy Perspectives* (blog), August 11, 2015, *https://blog.aee.net/understanding-irps-how-utilities-plan-for-the-future*.

[219] Dillon, "Small Utilities Say Subsidized Net-Metering."

[220] "State Net Metering Policies," National Conference of State Legislatures, November 20, 2017, *http://www.ncsl.org/research/energy/net-metering-policy-overview-and-state-legislative-updates.aspx*.

[221] "Net metering in Nevada," *Wikipedia*, updated November 1, 2019, *https://en.wikipedia.org/wiki/Net_metering_in_Nevada#AB_270_legislation*.

[222] Sean Whaley, "Worries over firearms surface in rooftop-solar debate," *Las Vegas Review-Journal*, February 10, 2016, *https://www.reviewjournal.com/business/energy/worries-over-firearms-surface-in-rooftop-solar-debate/*.

[223] Robert Walton, "Nevada governor signs net metering bill," *Utility Dive*, June 16, 2017, *https://www.utilitydive.com/news/nevada-governor-signs-net-metering-bill/445177/*.

[224] Sara Matasci, "California net metering: everything you need to know about NEM 2.0," *EnergySage* (website), January 2, 2019, *https://news.energysage.com/net-metering-2-0-in-california-everything-you-need-to-know/*.

[225] "Vermont: State Profile and Energy Estimates," *eia* (website of U.S. Energy Information Administration), upated July 18, 2019, *https://www.eia.gov/state/?sid=VT.*

[226] Figure taken from the web page "Consumer vs Prosumer: What's the Difference?" Office of Energy Efficiency & Renewable Energy of the Department of Energy, May 11, 2017, *https://www.energy.gov/eere/articles/consumer-vs-prosumer-whats-difference.*

[227] Christine Legere, "Falmouth neighbors want wind turbines removed," *Cape Cod Times*, September 22, 2018, *https://www.capecodtimes.com/news/20180921/falmouth-neighbors-want-wind-turbines-taken-down*

[228] Bill McKibben, "Power to the People," *The New Yorker*, June 22, 2015, *https://www.newyorker.com/magazine/2015/06/29/power-to-the-people.*

[229] Kenneth R. Rosen, "Inside the Haywire World of Beirut's Electricity Brokers," *Wired*, August 29, 2018, *https://www.wired.com/story/beruit-electricity-brokers/.*

[230] Robert Bryce, *A Question of Power: Electricity and the Wealth of Nations*, Public Affairs, March 10, 2020, *https://www.amazon.com/Question-Power-Electricity-Wealth-Nations/dp/1610397495.*

[231] Gayathri Vaidyanathan, "Coal Trumps Solar in India," *ClimateWire*, October 19, 2015; reprinted in *Scientific American*, *https://www.scientificamerican.com/article/coal-trumps-solar-in-india/#googDisableSync/.*

[232] Hans Rosling, "The magic washing machine" (video), TED, December 2010, *https://www.ted.com/talks/hans_rosling_and_the_magic_washing_machine?language=en.*

[233] *https://www.iso-ne.com/committees/industry-collaborations/consumer-liaison.*

[234] "Consumer Liaison Group Meeting Summary: December 6, 2018," ISO-NE, undated, *https://www.iso-ne.com/static-assets/documents/2018/12/clg_meeting_summary_december_6_2018_final.pdf.*

[235] "Consumer Liaison Group Meeting Summary: September 20, 2018," ISO-NE, undated, *https://www.iso-ne.com/static-assets/documents/2018/10/clg_meeting_summary_september_20_2018_final.pdf.*

[236] List of countries by electricity consumption, Wikipedia, updated April 6, 2020 *https://en.wikipedia.org/wiki/List_of_countries_by_electricity_consumption*

[237] Reuters, "Germany to phase out coal by 2038 in move away from fossil fuels," *CNBC* (website), January 26, 2019, *https://www.cnbc.com/2019/01/26/germany-to-phase-out-coal-by-2038-in-move-away-from-fossil-fuels.html.*

[238] Joshua S. Goldstein and Staffan A. Qvist, *A Bright Future: How Some Countries Have Solved Climate Change and the Rest Can Follow*," PublicAffairs, January 20. You can get this book from your favorite bookseller, or you can order from Amazon, *https://www.amazon.com/Bright-Future-Countries-Solved-Climate/dp/1541724100/ref=sr_1_1_twi_har_2?.*

[239] Robert Bryce, *A Question of Power,* chapter 7.

[240] Robert Bryce, *A Question of Power:* chapter 11.

[241] David JC MacKay, *Sustainable Energy — without the hot air* (UIT Cambridge Ltd., 2009), available at booksellers and online at *https://www.withouthotair.com.*

[242] MacKay, *Sustainable Energy*, chapter 8, "Hydroelectricity," 55-56. Can be read online at *https://www.withouthotair.com/c8/page_55.shtml.*

[243] For his story, see "David J. C. MacKay," *Wikipedia*, updated September 18, 2019, *https://en.wikipedia.org/wiki/David_J._C._MacKay.*

[244] MacKay, *Sustainable Energy*, 114. Can be read online at *https://www.withouthotair.com/c19/page_114.shtml.*

[245] Howard C. Shaffer and Meredith J. Angwin, "Vermont Electric Power in Transition," Coalition for Energy Solutions, April 2010, *http://www.coalitionforenergysolutions.org/vt_elec_pwr_in_transitionpr.pdf.*

[246] Goldstein and Qvist, *A Bright Future.*

[247] Geert De Clercq, "New EU rules aim to make power cheaper when the sun shines," Reuters News Agency, March 31, 2017, *https://af.reuters.com/article/idAFL5N1H835A?pageNumber=2&virtualBrandChannel=0.*

[248] "When you use electricity can make a difference in how much it costs (webpage)," *Alliant Energy*, undated, *https://www.alliantenergy.com/WaysToSave/SavingsTipsandPrograms/TimeofDayPricingIOWARES.*

[249] Gretchen Bakke, *The Grid: The Fraying Wires Between Americans and Our Energy Future*, Bloomsbury Publishing Plc, 2016.

[250] Stoltz, "'Smart' electric meters."

[251] Katherine Tweed, "Even With WiFi Interference, Meter Opt Out Not Popular," *gtm:* (website of Greentech Media), November 23, 2011, *https://www.greentechmedia.com/articles/read/even-with-wi-fi-interference-meter-opt-out-not-popular#gs.3l0aq5.*

[252] Katherine Tweed, "No Smart Meter, No Cost in Vermont," *gtm:* (website of Greentech Media), May 15, 2012, *https://www.greentechmedia.com/articles/read/no-smart-meter-no-cost-in-vermont#gs.3l0rjs%5C.*

[253] Tom Evslin, "Don't like smart meters? Opt out at your own expense," *VTDigger*, September 28, 2011, *https://vtdigger.org/2011/09/28/evslin-don't-like-smart-meters-opt-out-at-your-own-expense/.*

[254] Robert Walton, "Smart meter readings are a valid 'warrantless search,' court rules," *Utility Dive*, August 21, 2018, *https://www.utilitydive.com/news/smart-meter-readings-are-a-valid-warrantless-search-court-rules/530507/.*

[255] Peter Cappers, Annika Todd, and Greg Leventis, "Uses for Smart Meter Data Webinar Series," Electricity Markets & Policy Group (at Berkeley Lab), November 2018, *https://emp.lbl.gov/publications/uses-smart-meter-data-webinar-series.*

[256] Jeff St. John, "Tendril Merges With Simple Energy to Form Uplight," *gtm:* (website of Greentech Media), July 15, 2019, *https://www.greentechmedia.com/articles/read/tendril-simple-energy-merge-to-form-uplight.*

257 Marsha Stoltz, "'Smart' electric meters come to NJ, bringing fears of Big Brother," *northjersey.com*, May 29, 2018, *https://www.northjersey.com/story/news/new-jersey/2018/05/29/smart-meters-arrive-nj-bringing-fears-big-brother/577870002/.*

258 "Energy Star and the Connected, Smart Home (web page)," ICF International, Inc., September 15, 2017, *http://eedal2017.uci.edu/wp-content/uploads/Friday-30-Feldman.pdf*

259 "Residential Demand Response Programs (web page)," *ClearlyEnergy*, updated October 10, 2016, *https://www.clearlyenergy.com/residential-demand-response-programs.*

260 "Summer Discount Plan," *Southern California Edison* (website), undated, *https://www.sce.com/residential/rebates-savings/summer-discount-plan.*

261 Meredith Angwin, "The Oversold Smart Grid: Dismissing the Work of Women," *Yes Vermont Yankee*, April 7, 2013, *https://yesvy.blogspot.com/2013/04/the-over-sold-smart-grid-dismissing-work.html#.XeFxCKfMxTY.*

262 "Historical Data Graphs per Year: United States," *IndexMundi* (website), June 30, 2018, *https://www.indexmundi.com/g/g.aspx?c=us&v=81.*

263 Using the data from *IndexMundi* referenced in the previous note, we calculate:

In 2008, electricity consumption in the United States was 3,892 million MWh, population 304 million, for a usage of 12.8 MWh/person.

In 2017, electricity consumption was 3,913 million MWh, population 327 million, for a usage of 12.0 MWh/person.

An 0.8 MWh/person reduction in electricity usage is about 6%.

264 Bakke, *The Grid*, 154.

265 Both plants are described in the *Wikipedia* articles "Pilgrim Nuclear Power Station" and "Palisades Nuclear Generating Station," *https://en.wikipedia.org/wiki/Pilgrim_Nuclear_Power_Station* and *https://en.wikipedia.org/wiki/Palisades_Nuclear_Generating_Station.*

266 "Entergy Sells Natural Gas-Fired Power Plant in Rhode Island," *Power Engineering* (website), December 18, 2015, *https://www.power-eng.com/2015/12/18/entergy-sells-natural-gas-fired-power-plant-in-rhode-island/.*

267 Charles E. Jones, "FirstEnergy: Transforming to a Regulated Company," Edison Electric Institute, November 2016, *https://www.eei.org/about/meetings/Meeting_Documents/2016-11-FinConf-FirstEnergy.pdf.*

268 Will Davis, "November News," *ANS Nuclear Cafe* (website), November 16, 2016, *http://ansnuclearcafe.org/2016/11/16/november-news/#sthash.WU13uINZ.pRaGEJuC.dpbs.*

269 Meredith Angwin, "The Future of Nuclear in RTO Areas," *Yes Vermont Yankee*, November 21, 2016, *https://yesvy.blogspot.com/2016/11/the-future-of-nuclear-in-rto-areas.html#.XeF_eqfMxTZ.*

[270] Reid Frazier, "FirstEnergy says it's closing three nuclear plants; seeks federal help," StateImpact Pennsylvania, March 29, 2018, *https://stateimpact.npr.org/pennsylvania/2018/03/29/firstenergy-says-its-closing-three-nuclear-plants-seeks-federal-help/.*

[271] Iulia Gheorghiu, "Ohio House approves nuclear, coal subsidies, ditches renewables mandate," *Utility Dive*, May 30, 2019, *https://www.utilitydive.com/news/ohio-house-approves-nuclear-subsidies-bill-also-covering-2-coal-plants/555825/.*

[272] Sonal Patel, "FirstEnergy Suffers Steep Losses, Will Close Massive Coal Plant," *POWER* (website), February 21, 2018, *https://www.powermag.com/firstenergy-suffers-steep-losses-will-close-massive-coal-plant/.*

[273] Michael Kuser, "Mystic Closure Notice Leaves Room for Reversal," *RTO Insider*, April 1, 2018, *https://rtoinsider.com/exelon-iso-ne-mystic-generating-station-89508/.*

[274] Rich Heidorn Jr, "Exelon Bid to Keep Mystic Units Running Provokes Outrage," *RTO Insider*, May 7, 2020, *https://rtoinsider.com/exelon-to-keep-mystic-units-running-provokes-outrage-162263/*

[275] Bride, New England CLG presentation.

[276] Robert Hargraves, *Thorium: energy cheaper than coal*, CreateSpace Independent Publishing Platform, July 25, 2012. Listed with *goodreads* (website), *https://www.goodreads.com/book/show/16192018-thorium.* Figure used with permission.

[277] Hans Rosling, with Ala Rosling and Anna Rosling Ronnlund, *Factfulness: Ten Reasons We're Wrong About the World—and Why Things Are Better Than You Think*, Flatiron Books, 2018, *https://us.macmillan.com/books/9781250107817.*

[278] Robert Bryce, *A Question of Power,* chapters 5 and 6.

[279] Kavulla, "There Is No Free Market for Electricity."

[280] "Managing oversupply," *Calfornia ISO* (website), frequently updated, *http://www.caiso.com/informed/Pages/ManagingOversupply.aspx.* Figure is from the November 18, 2019 update.

[281] Robert Walton, "CAISO expansion legislation delayed until 2018," *Utility Dive*, September 14, 2017, *https://www.utilitydive.com/news/caiso-expansion-legislation-delayed-until-2018/504940/.*

[282] David Roberts, "California's huge energy decision: link its grid to its neighbors, or stay autonomous?" *Vox* (website of VoxMedia), August 23, 2018, *https://www.vox.com/energy-and-environment/2018/7/31/17611288/california-energy-grid-regionalization-caiso-wecc-iso.*

[283] Independent Electricity System Operator (ieso) website, http://www.ieso.ca

[284] "What is Global Adjustment?" ieso, *http://www.ieso.ca/en/learn/electricity-pricing/what-is-global-adjustment.*

[285] "2019 Year in Review," *ieso, http://www.ieso.ca/en/Corporate-IESO/Media/Year-End-Data*

[286] "What We Do," *ieso, http://www.ieso.ca/en/Learn/About-the-IESO/What-We-Do*

[287] "Archived—Ontario's Long-Term Energy Plan, *ieso,* 2017 , *https://www.ontario.ca/page/ontarios-long-term-energy-plan*

[288] "Global Adjustment (GA,)" *ieso, http://www.ieso.ca/en/Power-Data/Price-Overview/Global-Adjustment*

[289] "2019 Year in Review, *ieso.*

[290] "Feedback, Responses, and Analysis: Floor Price Focus Group," Report to SE-91 Renewables Integration meeting, January 24, 2012, Revised, page 9.

[291] "Retail Electricity Price Reform: Path to Lower Energy Bills and Economy-Wide CO_2 Reductions," Ontario Society of Professional Engineers, April 2019 (available by download), *https://ospe.on.ca/advocacy/our-work/research-reports/.*

[292] For this section, I owe great appreciation for guidance to Paul Acchione, P.E. Fellow of OSPE report, "Retail Electricity Price Reform." I also appreciate conversations with bloggers Steve Aplin (*Canadian Energy Issues http://canadianenergyissues.com*) and Scott Luft (*Cold Air Online http://coldair.luftonline.net*). Any errors, however, are fully my own.

INDEX

C

G

H

I

O

P

Q

R

T

U

ABOUT THE AUTHOR

As a working chemist, Meredith Angwin headed projects that lowered pollution and increased reliability on the electric grid. Her work included pollution control for nitrogen oxides in gas-fired combustion turbines, and corrosion control in geothermal and nuclear systems.

She was one of the first women to be a project manager at the Electric Power Research Institute. She led projects in renewable and nuclear energy.

In the past ten years, she began to study and take part in grid oversight and governance. For four years, she served on the Coordinating Committee for the Consumer Liaison Group associated with ISO-NE, her local grid operator. She teaches courses and presents workshops on the electric grid.

She is also an advocate for nuclear energy. Her previous major book was *Campaigning for Clean Air: Strategies for Pro-Nuclear Advocacy.*

(The audio book was a finalist for the Voice Arts award in the business category.) Meredith has been keynote or featured speaker at several nuclear events, including keynote at the worldwide Nuclear Science Week in 2018.

She and her husband George live in Vermont. They have two children and four grandchildren who live in the New York City area.

Contact her at *meredithangwin@gmail.com*

I hope you enjoyed this book. Can you do me a favor? Please post a review.

Like all authors, I rely on online reviews to encourage further sales. Your opinion is valuable, both to me and to other potential readers. Can you take a few minutes to review the book on Amazon or on any other book-review website that you prefer? Thank you very much!

Made in United States
Troutdale, OR
07/17/2023

11323997R00246